Topology

Topology

A Categorical Approach

Tai-Danae Bradley, Tyler Bryson, and John Terilla

The MIT Press
Cambridge, Massachusetts
London, England

This book was set in Times Roman and Computer Modern by the authors. Printed and bound in the United States of America.

Library of Congress Cataloging-in-Publication Data is available.

ISBN: 978-0-262-53935-7

10 9 8 7 6 5 4 3 2

Contents

Preface

When teaching a graduate topology course, it's tempting to rush through the point-set topology, or even skip it altogether, and do more algebraic topology, which is more fun to teach and more relevant to today's students. Many point-set topology ideas are already familiar to students from real analysis or undergraduate point-set topology courses and may seem safe to skip. Also, point-set ideas that might be unfamiliar but important in other subjects, say the Zariski topology in algebraic geometry or the p-adic topology in number theory, can be picked up later when they are encountered in context.

An alternative to rushing through point-set topology is to cover it from a more modern, categorical point of view. We think this alternative is better for several reasons. Since many students are familiar with point-set ideas already, they are in a good position to learn something new about these ideas, like the universal properties characterizing them. Plus, using categorical methods to handle point-set topology, whose name even suggests an old-fashioned way of thinking of spaces, demonstrates the power and versatility of the methods. The category of topological spaces is poorly behaved in some respects, but this provides opportunities to draw meaningful contrasts between topology and other subjects and to give good reasons why some kinds of spaces (like compactly generated weakly Hausdorff spaces) enjoy particular prevalence. Finally, there is the practicality that point-set topology is on the syllabus for our first-year topology courses and PhD exams. Teaching the material in a way that both deepens understanding and prepares a solid foundation for future work in modern mathematics is an excellent alternative.

This text contains material curated from many resources to present elementary topology from a categorical perspective. In particular, we cover some of the same topics as Ronnie Brown (2006), although our outlook is, from the outset, more categorical. The result is intentionally less comprehensive but more widely useful. We assume that students know linear algebra well and have had at least enough abstract algebra to understand how to form the quotient of a group by a normal subgroup. Students should also have some basic knowledge about how to work with sets and their elements, even as they endeavor to work with arrows instead. Students encountering diagrams and arrows for the first time may want to spend a little extra time reading the preliminaries where the objects (sets) are presumably familiar but the perspective may be new.

Covering spaces, homology, and cohomology are not in this book, but students will be ready to learn more algebraic topology after reading through our text. The omitted topics are likely included in whichever algebraic topology book is used afterward, including Massey (1991), Rotman (1998), May (1999), Hatcher (2002), and tom Dieck (2008), for which the reader will be well prepared. When we teach the first semester topology course in our PhD program, we usually cover the classification of compact surfaces. While this classification theorem is not in the text, an instructor may wish to cover it in their course, and it is hard to beat Conway's ZIP proof or the proof in Massey (1991).

With detailed descriptions of topological constructions emphasizing universal properties; filter-based treatment of convergence; thorough discussions of limits, colimits, and adjunctions; and an early emphasis on homotopy, this book guides the student of topology through the important transition from an undergraduate with a solid background in analysis or point-set topology to a graduate student preparing to work on problems in contemporary mathematics.

0 Preliminaries

I argue that set theory should not be based on membership, as in Zermelo-Frankel set theory, but rather on isomorphism-invariant structure.
—William Lawvere (Freitas, 2007)

Introduction. Traditionally, the first chapter of a textbook on mathematics begins by recalling basic notions from set theory. This chapter begins by introducing basic notions from category theory, the shift being from the internal anatomy of sets to their relationships with other sets. The idea of focusing on the relationships between mathematical objects, rather than on their internals, is fundamental to modern mathematics, and category theory is the framework for working from this perspective. Our goal for chapter 0 is to present what is perhaps familiar to you—functions, sets, topological spaces—from the contemporary perspective of category theory. Notably, category theory originated in topology in the 1940s with work of Samuel Eilenberg and Saunders MacLane (Eilenberg and MacLane, 1945).

This chapter's material is organized into three sections. Section 0.1 begins with a quick review of topological spaces, bases, and continuous functions. Motivated by a few key features of topological spaces and continuous functions, we'll proceed to section 0.2 and introduce three basic concepts of category theory: categories, functors, and natural transformations. The same section highlights one of the main philosophies of category theory, namely that studying a mathematical object is akin to studying its relationships to other objects. This golden thread starts in section 0.2 and weaves its way through the remaining pages of the book—we encourage you to keep an eye out for the occasional glimmer. Finally, equipped with the categorical mindset, we'll revisit some familiar ideas from basic set theory in section 0.3.

0.1 Basic Topology

Definition 0.1 A *topological space* $(X, \mathcal{T})$ consists of a set X and a collection $\mathcal{T}$ of subsets of X that satisfy the following properties:

(i) The empty set $\varnothing$ and X are in $\mathcal{T}$.
(ii) Any union of elements in $\mathcal{T}$ is also in $\mathcal{T}$.
(iii) Any finite intersection of elements in $\mathcal{T}$ is also in $\mathcal{T}$.

The collection $\mathcal{T}$ is called a *topology* on X, and we'll write X in place of $(X, \mathcal{T})$ if the topology is understood. Occasionally, we'll also refer to the topological space X as simply a *space*. Elements of the topology $\mathcal{T}$ are called *open sets*, and a set is called *closed* if and only if its complement is open.

Example 0.1 Suppose X is any set. The collection 2^X of all subsets of X forms a topology called the *discrete topology* on X, and the set $\{\varnothing, X\}$ forms a topology on X called the *indiscrete topology* or the *trivial topology*.

Sometimes, two topologies on the same set are comparable. When $\mathcal{T} \subseteq \mathcal{T}'$, the topology $\mathcal{T}$ can be called *coarser* than $\mathcal{T}'$, or the topology $\mathcal{T}'$ can be called *finer* than $\mathcal{T}$. Instead of coarser and finer, some people say "smaller and larger" or "weaker and stronger," but the terminology becomes clearer—as with most things in life—with coffee. A *coarse* grind yields a *small* number of chunky coffee pieces, whereas a *fine* grind results in a *large* number of tiny coffee pieces. Finely ground beans make stronger coffee; coarsely ground beans make weaker coffee.

In practice, it can be easier to work with a small collection of open subsets of X that generates the topology.

Definition 0.2 A collection $\mathcal{B}$ of subsets of a set X is a *basis* for a topology on X if and only if

(i) For each $x \in X$ there is a $B \in \mathcal{B}$ such that $x \in B$.

(ii) If $x \in A \cap B$ where $A, B \in \mathcal{B}$, then there is at least one $C \in \mathcal{B}$ such that $x \in C \subseteq A \cap B$.

The topology $\mathcal{T}$ *generated* by the basis $\mathcal{B}$ is defined to be the coarsest topology containing $\mathcal{B}$. Equivalently, a set $U \subseteq X$ is open in the topology generated by the basis $\mathcal{B}$ if and only if for every $x \in U$ there is a $B \in \mathcal{B}$ such that $x \in B \subseteq U$.

We call the collection of $B \in \mathcal{B}$ with $x \in B$ the *basic open neighborhoods* of x. More generally, in any topology $\mathcal{T}$, those $U \in \mathcal{T}$ containing x are called *open neighborhoods* of x and together are denoted $\mathcal{T}_x$.

Example 0.2 A *metric space* is a pair (X, d) where X is a set and $d\colon X \times X \to \mathbb{R}$ is a function satisfying

- $d(x, y) \geq 0$ for all $x, y \in X$,
- $d(x, y) = d(y, x)$ for all $x, y \in X$,
- $d(x, y) + d(y, z) \geq d(x, z)$ for all $x, y, z \in X$,
- $d(x, y) = 0$ if and only if $x = y$ for all $x, y \in X$.

The function d is called a *metric* or a *distance function*. If (X, d) is a metric space, $x \in X$, and $r > 0$, then the ball centered at x of radius r is defined to be

$$B(x, r) = \{y \in X \mid d(x, y) < r\}.$$

The balls $\{B(x,r)\}$ form a basis for a topology on X called *the metric topology*. So any set with a metric gives rise to a topological space. On the other hand, if Y is a space with topology $\mathcal{T}$ and if there is a metric d on Y such that the metric topology is the same as $\mathcal{T}$, then Y is said to be *metrizable*.

Any subset of a metric space is a metric space. In particular, subsets of $\mathbb{R}^n$ provide numerous examples of topological spaces since $\mathbb{R}^n$ with the usual Euclidean distance function is a metric space. For example,

- the real line $\mathbb{R}$,
- the unit interval $I := [0,1]$,
- the closed unit ball $D^n := \{(x_1,\ldots,x_n) \in \mathbb{R}^n \mid x_1^2 + \cdots + x_n^2 \leq 1\}$, and
- the *n-sphere* $S^n := \{(x_1,\ldots,x_{n+1}) \in \mathbb{R}^{n+1} \mid x_1^2 + \cdots + x_{n+1}^2 = 1\}$

are all important topological spaces.

We'll see more examples of topological spaces in chapter 1 and will discuss some important features in chapter 2. One way we'll glean information is by studying how spaces relate to each other. These relationships are best understood as functions that interact nicely—in the sense of the following definition—with open subsets of the spaces.

Definition 0.3 A function $f\colon X \to Y$ between two topological spaces is *continuous* if and only if $f^{-1}U$ is open in X whenever U is open in Y.

It is straightforward to check that for any topological space X, the identity $\mathrm{id}_X\colon X \to X$ is continuous, and for any topological spaces X, Y, Z and any continuous functions $f\colon X \to Y$ and $g\colon Y \to Z$, the composition $gf := g \circ f\colon X \to Z$ is continuous, and moreover that this composition is associative. On the surface, these observations may appear to be ho-hum, but quite the opposite is true. Collectively, the seemingly routine observations above amount to the statement that *topological spaces together with continuous functions form a category.*

0.2 Basic Category Theory

In this section, we'll give the formal definition of a *category* along with several examples.

0.2.1 Categories

Definition 0.4 A *category* C consists of the following data:

(i) a class of *objects*,

(ii) for every two objects X, Y, a set[1] $\mathsf{C}(X, Y)$ whose elements are called *morphisms* and denoted by arrows; for example $f\colon X \to Y$,

(iii) a *composition* rule defined for morphisms: if $f\colon X \to Y$ and $g\colon Y \to Z$, then there is a morphism $gf\colon X \to Z$.

These data must satisfy the following two conditions:

(i) Composition is associative. That is, if $h\colon X \to Y$, $g\colon Y \to Z$, $f\colon Z \to W$, then $f(gh) = (fg)h$.

(ii) There exist identity morphisms. That is, for every object X, there exists a morphism $\mathrm{id}_X\colon X \to X$ with the property that $f\,\mathrm{id}_X = f = \mathrm{id}_Y f$ whenever f is a morphism from X to Y. By the usual argument, identity morphisms are unique: if $\mathrm{id}'_X\colon X \to X$ is another identity morphism, then $\mathrm{id}'_X = \mathrm{id}'_X\,\mathrm{id}_X = \mathrm{id}_X$.

The associativity condition can also be expressed by way of a commutative diagram. A *diagram* can be thought of as a directed graph with morphisms as edges and with objects as vertices, though we'll give a more categorical definition in chapter 4. A diagram is said to *commute* (or is *commutative*) if all paths that share the same initial and final vertex are the same. For example, if $h\colon X \to Y$ and $g\colon Y \to Z$ are composable morphisms, then there is a commutative diagram:

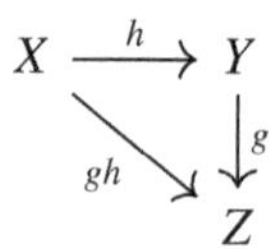

And if $f\colon Z \to W$ is a third morphism, then the property, "composition is associative,"—that is, $f(gh) = (fg)h$—is equivalent to the statement that the following diagram commutes:

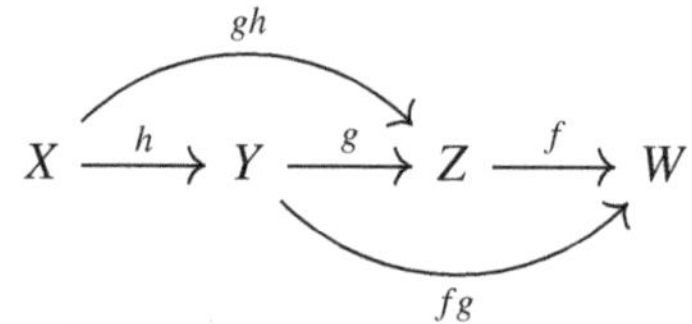

Simply put, a diagram is a visualization of morphism composition. A commutative diagram is a visualization of equalities between compositions. We'll see many more examples

[1] Some authors denote the set of morphisms from X to Y by $\hom_{\mathsf{C}}(X, Y)$ or simply $\hom(X, Y)$ if the category C is understood. Since keeping track of subscripts can be a chore, we'll usually promote the subscript to the forefront and write $\mathsf{C}(X, Y)$. While this notation may take some getting used to, we encourage the reader to do so. Authors also differ in their definitions of "category." We require a category to have *only a set's worth* of morphisms between any two objects—a property known as being *locally small*. Others sometimes allow categories to have more arrows.

in the pages to come. And speaking of examples, we already mentioned that topological spaces with continuous functions form a category. Here is another.

Example 0.3 For any given field $\mathbf{k}$, there is a category denoted $\mathsf{Vect}_{\mathbf{k}}$ whose objects $V, W, \ldots$ are vector spaces over $\mathbf{k}$ and whose morphisms are linear transformations. To verify the claim, suppose $T: V \to W$ and $S: W \to U$ are linear transformations. Then for any $v, v' \in V$ and any $k \in \mathbf{k}$,

$$ST(kv + v') = S(kTv + Tv') = kSTv + STv'$$

and so $ST: V \to U$ is indeed a linear transformation. Associativity of composition is automatic since linear transformations are functions and composition of functions is always associative. And for any vector space, the identity function is a linear transformation. More generally, modules over a fixed ring R together with R-module homomorphisms form a category, $R\mathsf{Mod}$.

Here are a few more examples. This time we'll leave the verifications as an exercise.

- Set: The objects are sets, the morphisms are functions, and composition is composition of functions.
- Set_*: The objects are sets S having a distinguished element. (Such sets are called *pointed sets*.) A morphism $f: S \to T$ is a function satisfying $fs_0 = t_0$ whenever s_0 is the distinguished element of S and t_0 is the distinguished element of T. (Such functions are said to "respect" the distinguished elements.) Composition is composition of functions.
- Top: The objects are topological spaces, the morphisms are continuous functions, and composition is composition of functions.
- Top_*: The objects are topological spaces with a distinguished point, often called a basepoint. (Such spaces are called *pointed* or *based* spaces.) The morphisms are continuous functions that map basepoint to basepoint, and composition is composition of functions.
- hTop: The objects are topological spaces, the morphisms are homotopy classes of continuous functions, and composition is composition of these homotopy classes. The notion of homotopy will be introduced in section 1.6.
- Grp: The objects are groups, the morphisms are group homomorphisms, and composition is composition of homomorphisms.
- Every group G can be viewed as a category with one object $\bullet$ and with a morphism $g: \bullet \to \bullet$ for each group element g. The composition of two morphisms f, g corresponds to the group element gf.

- A directed multigraph determines a category whose objects are the vertices and whose morphisms are the directed paths along finitely many arrows joined head-to-tail.[2] For example, the directed graph

$$\bullet \longrightarrow \bullet \longrightarrow \bullet \longrightarrow \cdots$$

  determines a category that will make an appearance in example 4.1. For simple graphs, one might want to display more information by drawing identities and composite morphisms. To avoid confusion, one can state whether identities and composites are displayed. Here, of course, they are not drawn. This example also illustrates another notational convenience: we'll frequently use symbols such as $\bullet$ and $\circ$ as anonymous placeholders. So unless otherwise indicated, each "$\bullet$" should be considered a distinct object.
- For any category C, there is an opposite category C^{op}. The objects are the same as the objects of C, but the morphisms are reversed. Composition in C^{op} is defined by composition in C, that is, $\mathsf{C}^{\mathrm{op}}(X, Y) = \mathsf{C}(Y, X)$. To check that composition makes sense, suppose $f \in \mathsf{C}^{\mathrm{op}}(X, Y)$ and $g \in \mathsf{C}^{\mathrm{op}}(Y, Z)$ so that $f\colon Y \to X$ and $g\colon Z \to Y$. Then $fg\colon Z \to X$ and therefore $fg \in \mathsf{C}^{\mathrm{op}}(X, Z)$ as needed.

Category theory is an appropriate setting in which to discuss an age-old question: "When are two objects really the same?" The concept of "sameness" being a special kind of relationship means that objects are the same if there is a particular morphism—an *isomorphism*—between them. How we talk about sameness is at the heart of category theory. We take Wittgenstein's (1922) critique,

> Roughly speaking, to say of two things that they are identical is nonsense, and to say of one thing that it is identical with itself is to say nothing at all

as an invitation to ask for less, namely isomorphic (or uniquely isomorphic) objects, rather than identical objects.

Definition 0.5 Let X and Y be objects in any category, and suppose $f\colon X \to Y$.

(i) f is *left invertible* if and only if there exists a morphism $g\colon Y \to X$ so that $gf = \mathrm{id}_X$. The morphism g is called a *left inverse* of f.

(ii) f is *right invertible* if and only if there exists a morphism $h\colon Y \to X$ so that $fh = \mathrm{id}_Y$. The morphism h is called a *right inverse* of f.

In the case when f has both a left inverse g and a right inverse h, then

$$g = g\,\mathrm{id}_Y = gfh = \mathrm{id}_X\,h = h$$

and the single morphism $g = h$ is called the *inverse* of f. (We encourage you to verify our use of "the." If f has an inverse, it is unique.) Therefore

[2] Including—for each object—a unique path of length zero starting and ending at that object.

(iii) f is *invertible* and is said to be an *isomorphism* if it is both left and right invertible. Two objects X and Y are *isomorphic*, denoted $X \cong Y$, if there exists an isomorphism $f: X \to Y$.

The notion of *being isomorphic* is a form of equivalence, meaning it is reflexive, symmetric, and transitive—properties you should check. Isomorphic objects, therefore, form equivalence classes. Some categories have their own special terminology for both *isomorphism* and these *isomorphism classes*. For instance,

- Isomorphisms in Set are called *bijections*, and two isomorphic sets are said to have the same *cardinality*. A *cardinal* is an isomorphism class of sets.
- Isomorphisms in Top are called *homeomorphisms*, and two isomorphic spaces are said to be *homeomorphic*.
- Isomorphisms in hTop are called *homotopy equivalences*, and isomorphic spaces in hTop are said to be *homotopic*. (See section 1.6 for a discussion of homotopy.)

Mathematics is particularly concerned with properties that are preserved under isomorphisms in a given category. For instance, topology is essentially the study of properties that are preserved under homeomorphisms. Such properties are referred to as *topological properties* and distinguish spaces: if X and Y are homeomorphic and X has (or doesn't have) a certain property, then Y must have (or cannot have) that property, too.

Example 0.4 The cardinality of a topological space is a topological property since any homeomorphism $f: X \to Y$ is necessarily a bijective function, and so X and Y, when viewed as sets, must have the same cardinality. *Metrizability* is also a topological property. *Connectedness* (section 2.1), *compactness* (section 2.3), *Hausdorff* (section 2.2), *first countability* (section 3.2), . . . are examples of other topological properties that will be discussed in the indicated sections.

However, not every familiar property is a topological one.

Example 0.5 A metric space is called *complete* if every Cauchy sequence converges. Being a complete metric space is *not* a topological property. For instance, the map $(-1, 1) \to \mathbb{R}$ by $x \mapsto \frac{x}{(1-x^2)}$ is a homeomorphism, yet $\mathbb{R}$ is a complete metric space while $(-1, 1)$ is not. This example also shows that being bounded is also not a topological property: a metric space is said to be *bounded* if the metric is a bounded function. Clearly, $(-1, 1)$ is bounded while $\mathbb{R}$ is not.

Comparing an object X with another object can help us to understand X better. Taking this idea further, we may also compare X with *all* objects at once. In other words, we can learn a great deal of information about X by looking at all morphisms both *out of* and *in to* X. This is the content of the next theorem. In essence, it states that the isomorphism class of an object is completely determined by morphisms to and from it. This is one of the main maxims of category theory:

an object is completely determined by its relationships with other objects.

In fact, it's a corollary of a major result to be discussed in section 0.2.3. But before we state the theorem formally, here is some useful terminology.

Definition 0.6 For each morphism $f\colon X \to Y$ and object Z in a category, there is a map of sets $f_*\colon \mathsf{C}(Z,X) \to \mathsf{C}(Z,Y)$ called the *pushforward* of f defined by postcomposition $f_*\colon g \mapsto fg$.

$$\mathsf{C}(Z,X) \xrightarrow{f_*} \mathsf{C}(Z,Y)$$

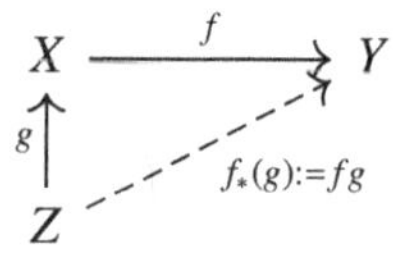

$$\begin{array}{ccc} X & \xrightarrow{f} & Y \\ {\scriptstyle g}\uparrow & \nearrow_{f_*(g):=fg} & \\ Z & & \end{array}$$

There is also a map of sets $f^*\colon \mathsf{C}(Y,Z) \to \mathsf{C}(X,Z)$ called the *pullback* defined by precomposition $f^*\colon g \mapsto gf$.

$$\mathsf{C}(X,Z) \xleftarrow{f^*} \mathsf{C}(Y,Z)$$

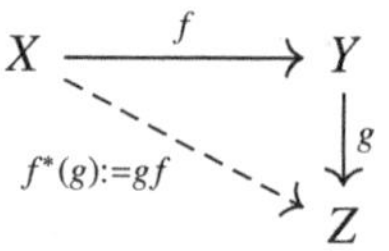

$$\begin{array}{ccc} X & \xrightarrow{f} & Y \\ & \searrow^{f^*(g):=gf} & \downarrow{\scriptstyle g} \\ & & Z \end{array}$$

The reason for the visual names *pushforward* and *pullback* is hopefully clear from the diagrams above. With this terminology in hand, here is the theorem whose summary we provided above.

Theorem 0.1 The following are equivalent.

- $f\colon X \to Y$ is an isomorphism.
- For every object Z, the pushforward $f_*\colon \mathsf{C}(Z,X) \to \mathsf{C}(Z,Y)$ is an isomorphism of sets.
- For every object Z, the pullback $f^*\colon \mathsf{C}(Y,Z) \to \mathsf{C}(X,Z)$ is an isomorphism of sets.

Proof. We'll prove that a morphism $f\colon X \to Y$ is an isomorphism if and only if for every object Z, the pushforward $f_*\colon \mathsf{C}(Z,X) \to \mathsf{C}(Z,Y)$ is an isomorphism of sets; then, we'll leave the other statements as exercises.

Suppose $f\colon X \to Y$ is an isomorphism. Let $g\colon Y \to X$ be the inverse of f. Then for any Z, the map $g_*\colon \mathsf{C}(Z,Y) \to \mathsf{C}(Z,X)$ is the inverse of f_*.

Conversely, suppose that for any Z, the map $f_*\colon \mathsf{C}(Z,X) \to \mathsf{C}(Z,Y)$ is an isomorphism of sets. Choosing $Z = Y$, we have $f_*\colon \mathsf{C}(Y,X) \xrightarrow{\cong} \mathsf{C}(Y,Y)$, so in particular f_* is surjective. Therefore there exists a morphism $g\colon Y \to X$ so that $f_*g = \mathrm{id}_Y$, which by definition implies $fg = \mathrm{id}_Y$. To see that $gf = \mathrm{id}_X$, consider the case $Z = X$. We know $f_*\colon \mathsf{C}(X,X) \xrightarrow{\cong} \mathsf{C}(X,Y)$, so in particular f_* is injective. And since $f_*(\mathrm{id}_X) = f$ and $f_*(gf) = fgf = f$, injectivity of f_* implies $\mathrm{id}_X = gf$ as needed. □

In summary, if you understand all the morphisms $X \to Z$, then you know X up to isomorphism. Or if you understand all the morphisms $Z \to X$, then you know X up to isomorphism.

Now it turns out that categories themselves are objects worthy of study. And as per our maxim above, studying a category should amount to studying its relationships to other categories. But what is a relationship between categories? It is called a *functor*. That is the subject of the next section.

You might wonder if we are about to construct a *category of categories* where the objects are categories and the morphisms are functors. This can be done, but there are a few important considerations. One consideration is size. When we defined categories, we assumed that for any two objects X and Y, there is a *set* of morphisms $\mathsf{C}(X, Y)$. This is convenient so that, as in the case of the pushforward $f_*\colon \mathsf{C}(Z, X) \to \mathsf{C}(Z, Y)$ of a morphism $f\colon X \to Y$ in the previous theorem, we can consider functions between sets of morphisms. However, if C and D are categories, then there may be *more* than a set's worth of functors $\mathsf{C} \to \mathsf{D}$. There's also another, more subtle consideration to think about: The sharpest identification of an object generally available when doing category is *up to unique isomorphism*. This should apply to the morphisms between categories as well. It can be dizzying, but the takeaway is that invertible functors provide a notion of equivalence that's too rigid to be useful. So instead of an "isomorphism of categories," we talk about an "equivalence of categories," which is defined in a more subtle way. A third consideration is that one might want to remember that categories themselves have objects and morphisms, and the category of categories has more structure than a category—it's probably better to think of it as a "higher" category, a concept we won't stop to discuss here.

0.2.2 Functors

Definition 0.7 A *functor* F from a category C to a category D consists of the following data:

(i) An object FX of the category D for each object X in the category C,
(ii) A morphism $Ff\colon FX \to FY$ for every morphism $f\colon X \to Y$.

These data must be compatible with composition and identity morphisms in the following sense:

(iii) $(Fg)(Ff) = F(gf)$ for any morphisms $f\colon X \to Y$ and $g\colon Y \to Z$,
(iv) $F\,\mathrm{id}_X = \mathrm{id}_{FX}$ for any object X.

A functor as defined above is sometimes described as a *covariant* functor to distinguish it from its *contravariant* counterpart. A functor $F\colon \mathsf{C}^{\mathrm{op}} \to \mathsf{D}$ whose domain is the opposite category is called a *contravariant* functor from C to D. Contravariant functors "reverse arrows." That is, for every morphism $f\colon X \to Y$, the functor $F\colon \mathsf{C}^{\mathrm{op}} \to \mathsf{D}$ assigns a morphism $Ff\colon FY \to FX$. We sometimes abuse terminology and refer to a contravariant functor simply as a functor. This usually doesn't cause confusion since the direction of the arrows can often be worked out easily.

Example 0.6 Here are some examples of functors.

- For an object X in a category C, there is a functor $\mathsf{C}(X,-)$ from C to Set that assigns to each object Z the set $\mathsf{C}(X,Z)$ and to each morphism $f\colon Y \to Z$ the pushforward f_* of f as in definition 0.6.

$$\begin{array}{ccc} Y & & \mathsf{C}(X,Y) \\ f\big\downarrow & \mapsto & \big\downarrow f_* \\ Z & & \mathsf{C}(X,Z) \end{array}$$

- For an object X in a category C, there is a functor $\mathsf{C}(-,X)$ from C^{op} to Set that assigns to each object Z the set $\mathsf{C}(Z,X)$ and to each morphism $f\colon Y \to Z$ the pullback f^* of f as in definition 0.6.

$$\begin{array}{ccc} Y & & \mathsf{C}(Y,X) \\ f\big\downarrow & \mapsto & \big\uparrow f^* \\ Z & & \mathsf{C}(Z,X) \end{array}$$

- Fix a set X. There is a functor $X\times -$ from Set to Set defined on objects by $Y \mapsto X\times Y$ and on morphisms f by $\mathrm{id}\times f$. As we'll see in example 5.1, the functors $X \times -$ and $\mathsf{C}(X,-)$ form a special pair called the *product-hom adjunction.*
- Fix a vector space V over a field $\mathbf{k}$. There is a functor $V\otimes -$ from $\mathsf{Vect}_{\mathbf{k}}$ to $\mathsf{Vect}_{\mathbf{k}}$ defined on objects by $W \mapsto V\otimes W$ and on morphisms by $f \mapsto \mathrm{id}\otimes f$.
- There is a *forgetful functor*, usually denoted U for "underlying," from Grp to Set that forgets the group operation. Concretely, it sends a group G to its underlying set UG and a group homomorphism to its underlying function.
- There is a *free functor* F from Set to Grp that assigns the free group FS to the set S. The free and forgetful functors form a special pair called the *free-forgetful adjunction*. We'll expand on this in section 5.2.
- There are other "forgetful" functors besides the one that forgets the group operation. Any functor that forgets structure, such as the $U : \mathsf{Top} \to \mathsf{Set}$ that forgets the topology, may be referred to as a forgetful functor.
- The *fundamental group*, to be discussed in detail in chapter 6, defines a functor π_1 from Top_* to Grp.
- The construction of the Grothendieck group of a commutative monoid is functorial. That is, there is a "Grothendieck group" functor from the category of commutative

monoids to the category of commutative groups that constructs a group from a commutative monoid by attaching inverses.

Here are a few different kinds of functors.

Definition 0.8 Let F be a functor from a category C to a category D. If for all objects X and Y in C, the map

$$\mathsf{C}(X, Y) \to \mathsf{D}(FX, FY) \qquad \text{given by} \qquad f \mapsto Ff$$

(i) is injective, then F is called *faithful*;
(ii) is surjective, then F is called *full*;
(iii) is bijective, then F is called *fully faithful*.

Fully faithful functors preserve all relationships among objects of the domain category: F is fully faithful if and only if each $FX \to FY$ is the image of *exactly one* morphism $X \to Y$. Think of it as an embedding, so to speak, of one category into another. But note, a fully faithful functor need not be injective on objects, so we use the term "embedding" loosely here. A fully faithful functor that *is* injective on objects is called a *full embedding*.

We'll define one of category theory's most famous fully faithful functors in the following section, but first we emphasize an important utility of functors. A functor $F: \mathsf{C} \to \mathsf{D}$ encodes invariants of isomorphism classes of objects within C. This is a fundamental idea, arising from the fact that functors take compositions to compositions and identities to identities. Consequentially,

functors take isomorphisms to isomorphisms.

So if two objects X and Y are "the same" in C, then FX and FY must be "the same" in D, the contrapositive being just as useful. For instance, the value that any functor $\mathsf{Top} \to \mathsf{C}$ assigns to a topological space is automatically a topological property (or we might say *a topological invariant*). So the question arises, "What is a useful choice of category C?" If we choose C to be an algebraic category, then we enter into the realm of *algebraic topology*. A *homology theory*, for example, is a functor $H: \mathsf{Top} \to R\mathsf{Mod}$ and is an excellent means of distinguishing topological spaces: if HX and HY are not isomorphic R-modules, then X and Y are not isomorphic spaces.

Having adopted the perspective that functors are invariants, it's natural to wonder: "When are two invariants the same?" To answer this, one needs a way to compare functors.

0.2.3 Natural Transformations and the Yoneda Lemma

Definition 0.9 Let F and G be functors $\mathsf{C} \to \mathsf{D}$. A *natural transformation* η from F to G consists of a morphism $\eta_X: FX \to GX$ for each object X in C. Moreover, these morphisms in D must satisfy the property that $\eta_Y Ff = Gf\eta_X$ for every morphism $f: X \to Y$ in C. In other words, the following diagram must commute:

$$\begin{array}{ccc} FX & \xrightarrow{Ff} & FY \\ {\scriptstyle \eta_X}\downarrow & & \downarrow{\scriptstyle \eta_Y} \\ GX & \xrightarrow{Gf} & GY \end{array}$$

For any two functors $F, G\colon \mathsf{C} \to \mathsf{D}$, let $\mathsf{Nat}(F, G)$ denote the natural transformations from F to G. If $\eta_X\colon FX \xrightarrow{\cong} GX$ is an isomorphism for each X, then η is called a *natural isomorphism* or a *natural equivalence*, and we say F and G are *naturally isomorphic*, denoted $F \cong G$.

Keep in mind that a natural transformation η is the *totality* of all the morphisms η_X, and each η_X can be thought of as a component, so to speak, of η. Simply put, a natural transformation is a collection of maps from one diagram to another, and these maps are special in that they *commute* with all the arrows in the diagrams.[3]

The language of natural transformations not only provides us an avenue in which comparing invariants (functors) becomes possible, but it also prompts us to revisit an idea mentioned in section 0.2.1. There we introduced the categorical philosophy that studying a mathematical object is more of a global, as opposed to a local, endeavor. That is, we can paint a better—rather, *a complete*—picture of an object once we investigate its interactions with all other objects. This theme finds its origins in an—if not *the most*—important result in category theory.

Yoneda Lemma For every object X in C and for every functor $F\colon \mathsf{C}^{\mathrm{op}} \to \mathsf{Set}$, the set of natural transformations from $\mathsf{C}(-, X)$ to F is isomorphic to FX,

$$\mathsf{Nat}(\mathsf{C}(-, X), F) \cong FX$$

In other words, elements of the set FX are in bijection with natural transformations from $\mathsf{C}(-, X)$ to F. We omit the proof but take note of the special case when $F = \mathsf{C}(-, Y)$:

$$\mathsf{Nat}(\mathsf{C}(-, X), \mathsf{C}(-, Y)) \cong \mathsf{C}(X, Y) \tag{0.1}$$

It's closely related to theorem 0.1 in the following way.

First observe that given two categories C and D, we can form a new category D^{C} whose objects are functors $\mathsf{C} \to \mathsf{D}$ and whose morphisms are natural transformations. To ensure that D^{C} is locally small, we may require C to be small and D to be locally small. When considering contravariant functors and taking $D = \mathsf{Set}$, we obtain the category $\mathsf{Set}^{\mathsf{C}^{\mathrm{op}}}$. An object here is called a *presheaf*. This is a very nice category—it has all finite limits and

[3] So a natural transformation may be viewed simultaneously as a single arrow between two functors or as a collection of arrows between two diagrams. Therefore you might hope that diagrams can be viewed as functors. As we will see in section 4.1, this is indeed the case.

colimits (to be discussed in chapter 4), it is Cartesian closed, and it forms what's called a topos. We won't dwell on these properties, though. Instead we turn our attention to a special functor $y\colon \mathsf{C} \to \mathsf{Set}^{\mathsf{C}^{\mathrm{op}}}$. It's defined by sending an object X to the presheaf $\mathsf{C}(-,X)$ and a morphism $f\colon X \to Y$ to the natural transformation f_*,

$$\begin{array}{ccc} X & & \mathsf{C}(-,X) \\ {\scriptstyle f}\downarrow & \longmapsto & \downarrow{\scriptstyle f_*} \\ Y & & \mathsf{C}(-,Y) \end{array}$$

This is a slight abuse of notation, though. By $f_*\colon \mathsf{C}(-,X) \to \mathsf{C}(-,Y)$ we mean the natural transformation whose component morphisms are the pushforward of f.

The isomorphism in (0.1) shows that the functor y is fully faithful and therefore *embeds* C into the functor category $\mathsf{Set}^{\mathsf{C}^{\mathrm{op}}}$. For this reason, y is called the *Yoneda embedding*. The punchline is that each object X in C can be viewed as the contravariant functor $\mathsf{C}(-,X)$. Practically speaking, this means information about X can be obtained by studying the set of all morphisms in to it. But what about morphisms *out of* it?

There is a version of the Yoneda lemma for covariant functors (called *co-presheaves*) in the category $\mathsf{Set}^{\mathsf{C}}$ and, accordingly, a contravariant Yoneda embedding $\mathsf{C}^{\mathrm{op}} \to \mathsf{Set}^{\mathsf{C}}$. The corresponding result is that X can also be viewed as a functor $\mathsf{C}(X,-)$. So the moral of the story is that if you understand maps in and out of X, then you understand X. Thus we've come full circle to the theme introduced in section 0.2.1:

objects are completely determined by their relationships with other objects.

The adjective "completely" is justified by the following important corollary of the Yoneda lemma:

$$X \cong Y \qquad \text{if and only if} \qquad \mathsf{C}(-,X) \cong \mathsf{C}(-,Y).$$

One direction follows from the fact that the Yoneda embedding is a *functor*. The other direction follows from the fact that it is fully faithful. It's also true that $X \cong Y$ if and only if $\mathsf{C}(X,-) \cong \mathsf{C}(Y,-)$, as can be verified by considering the contravariant Yoneda embedding. And with that, we've just provided a restatement of theorem 0.1.

As a final remark, at times we'll adapt this philosophy and use it to consider just a few (and not necessarily all) morphisms to or from an object—this also yields fruitful information.

Example 0.7 Two sets X and Y are isomorphic if and only if $\mathsf{Set}(Z,X) \cong \mathsf{Set}(Z,Y)$ for all sets Z. That is, X and Y are the same if and only if they relate to all other sets in the same way. But this is overkill! Two sets are isomorphic if and only if they have the same cardinality, so to distinguish X and Y we need only look at the case when Z is the one-point set $*$. Indeed, a morphism $* \to X$ is a choice of element $x \in X$, and X and Y will have the same cardinality if and only if $\mathsf{Set}(*,X) \cong \mathsf{Set}(*,Y)$.

Perhaps it's not surprising that the full arsenal of the Yoneda lemma is not needed to distinguish sets. After all, they have no internal structure. The strength of lemma is more clearly seen when we look at objects such as groups or topological spaces that have more interesting aspects. But we include the example to draw attention to some notation you'll likely have noticed: we choose to omit the parentheses around the arguments of morphisms, for example, fx versus $f(x)$. This is consistent with category theory's frequent emphasis on morphisms, for as we've just seen, every element x in a set X can be viewed as a morphism $x\colon * \to X$. So given a function $f\colon X \to Y$, the image of x in Y is ultimately the composite morphism fx.

Keeping in tune with the previous paragraph, we now proceed to a discussion on basic set theory by recasting familiar material in a more categorical light.

0.3 Basic Set Theory

We'll begin with a brief review of functions.

0.3.1 Functions

A function is said to be *injective* if and only if it is *left cancellative.* That is, $f\colon X \to Y$ is injective if and only if for all functions $g_1, g_2\colon Z \to X$ with $fg_1 = fg_2$, it follows that $g_1 = g_2$. Equivalently, f is injective if and only if $f_*\colon \mathsf{Set}(Z, X) \to \mathsf{Set}(Z, Y)$ is injective for all Z. Yet another equivalent definition is that f is injective if and only if it has a left inverse, that is, if and only if there exists $g\colon Y \to X$ so that $gf = \mathrm{id}_Y$. Note that the composition of injective functions is injective; also, for any $f\colon X \to Y$ and $g\colon Y \to Z$, if gf is injective, then f is injective. We will denote injective functions by hooked arrows, as in $f\colon X \hookrightarrow Y$.

More generally, left-cancellative morphisms in any category are called *monomorphisms* or said to be *monic* and are denoted with arrows with tails as in $X \rightarrowtail Y$. In this more general case, left invertible implies left cancellative, but not conversely. For example, the map $n \mapsto 2n$ defines a left-cancellative group homomorphism $f\colon \mathbb{Z}/2\mathbb{Z} \to \mathbb{Z}/4\mathbb{Z}$. However, there is no group homomorphism $g\colon \mathbb{Z}/4\mathbb{Z} \to \mathbb{Z}/2\mathbb{Z}$ so that $gf = \mathrm{id}_{\mathbb{Z}/2\mathbb{Z}}$.

A function is said to be *surjective* if and only if it is *right cancellative.* That is, $f\colon X \to Y$ is *surjective* if and only if for all functions $g_1, g_2\colon Y \to Z$ with $g_1 f = g_2 f$, it follows that $g_1 = g_2$. Equivalently, f is surjective if and only if $f^*\colon \mathsf{Set}(Y, Z) \to \mathsf{Set}(X, Z)$ is injective for all Z or, equivalent still, if and only if it has a right inverse. That is, $f\colon X \to Y$ is surjective if and only if there exists $g\colon Y \to X$ so that $fg = \mathrm{id}_X$. The composition of surjective functions is surjective, and for any $f\colon X \to Y$ and $g\colon Y \to Z$, if gf is surjective, then g is surjective. Surjective functions will be denoted with two headed arrows as in $f\colon X \twoheadrightarrow Y$.

In general, right-cancellative morphisms in any category are called *epimorphisms* or said to be *epic* and are also denoted with two-headed arrows. Right invertible implies right cancellative in any category, but not conversely. (You are asked to provide an example in exercise 0.3 at the end of the chapter.)

Finally, note that in Set a function that is both injective and surjective is an isomorphism. This is because left invertible and right invertible imply invertible. (You should check that having a left inverse and a right inverse both imply there's a single two-sided inverse). But left cancellative and right cancellative together do not imply invertible: there are categories—and Top is one of them—that have morphisms that are both monic and epic, which nonetheless fail to be isomorphisms.

0.3.2 The Empty Set and One-Point Set

The empty set $\varnothing$ is *initial* in Set. That is, for any set X there is a unique function $\varnothing \to X$. On the other hand, the one-point set $*$ is *terminal*. That is, for any set X, there is a unique function $X \to *$. You might take issue with the definite article "the" in "the one-point set," but it is standard to use the definite article in circumstances that are *unique up to unique isomorphism*. That is the case here: if $*$ and $*'$ are both one-point sets, then there is a unique isomorphism $* \xrightarrow{\cong} *'$. Note also that this terminology is not unique to Set: we can make sense of initial and terminal objects in any category, though such objects may not always exist. An object C in a category C is called terminal if for every object X in C there is a unique morphism $X \to C$. Dually, an object D is called initial if for every object X in C there is a unique morphism $D \to X$.

0.3.3 Products and Coproducts in Set

The *Cartesian product* of two sets X and Y is defined to be the set $X \times Y$ of all ordered pairs (x, y) such that $x \in X$ and $y \in Y$. While this tells us what the set $X \times Y$ *is*, it doesn't say much about the properties that it possesses or how it relates to other sets in the category. This prompts us to look for a more categorical description of the Cartesian product.

The Cartesian product of two sets X and Y is a set $X \times Y$ that comes with maps $\pi_1 : X \times Y \to X$ and $\pi_2 : X \times Y \to Y$. The product is characterized by the property that for any set Z and any functions $f_1 : Z \to X$ and $f_2 : Z \to Y$, there is a unique map $h : Z \to X \times Y$ with $\pi_1 h = f_1$ and $\pi_2 h = f_2$. By "characterized by," we mean that the Cartesian product is the *unique* (up to isomorphism) object in Set with this property.

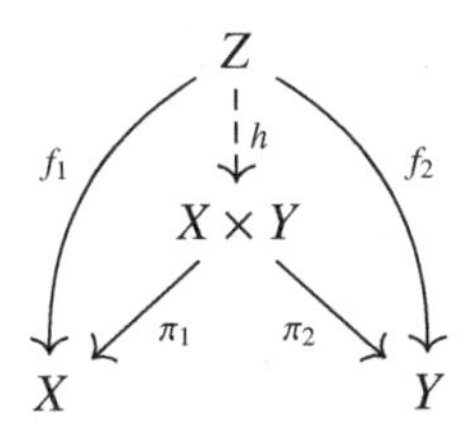

As an example, note the isomorphism of finite sets $\{1, \dots, n\} \times \{1, \dots, m\} \cong \{1, \dots, nm\}$.

The *disjoint union* of two sets X and Y also has a categorical description. It is a set $X \coprod Y$ that comes with maps $i_1 : X \to X \coprod Y$ and $i_2 : Y \to X \coprod Y$ and is characterized by the property that for any set Z and any functions $f_1 : X \to Z$ and $f_2 : Y \to Z$, there is a unique map $h: X \coprod Y \to Z$ with $hi_1 = f_1$ and $hi_2 = f_2$.

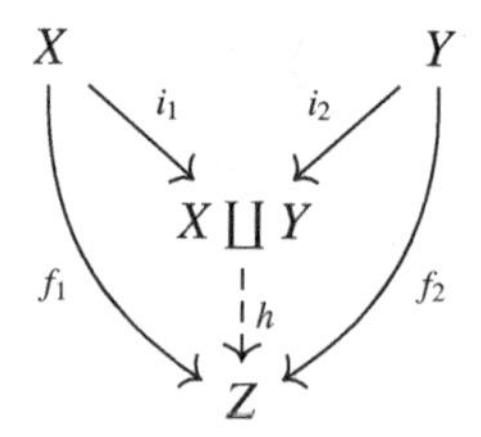

Sometimes, disjoint union is called the sum and is denoted $X + Y$ or $X \oplus Y$ instead of $X \coprod Y$. As an example, note that $\{1, \ldots, n\} + \{1, \ldots, m\} \cong \{1, \ldots, n + m\}$. The property characterizing the disjoint union is dual to the one characterizing the product, and the disjoint union is sometimes called the *coproduct* of sets.

We can also take products and disjoint unions of arbitrary collections of sets. The disjoint union of a collection of sets $\{X_\alpha\}_{\alpha \in A}$ is a set $\coprod_{\alpha \in A} X_\alpha$ together with maps $i_\alpha : X_\alpha \to \coprod X_\alpha$ satisfying the property that for any set Z and any collection of functions $\{f_\alpha : X_\alpha \to Z\}$, there is a unique map $h: \coprod X_\alpha \to Z$ with $hi_\alpha = f_\alpha$ for all $\alpha \in A$.

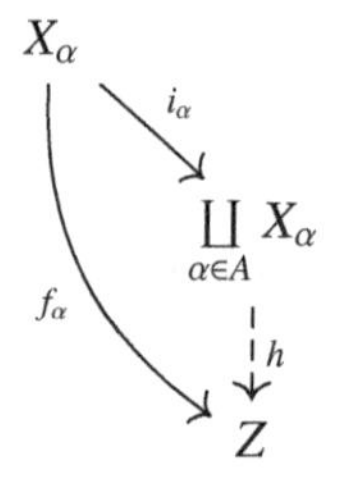

The product of a collection of sets $\{X_\alpha\}_{\alpha \in A}$ is sometimes described as the subset of functions $f: A \to \coprod X_\alpha$ satisfying $f(\alpha) \in X_\alpha$. But, as we mentioned earlier, what's more important than the construction of the product is to understand its universal property. The product of a collection of sets $\{X_\alpha\}_{\alpha \in A}$ is a set $\prod_{\alpha \in A} X_\alpha$ together with maps $\pi_\alpha : \prod X_\alpha \to X_\alpha$ characterized by the property that for any set Z and any collection of functions $\{f_\alpha : Z \to X_\alpha\}$, there is a unique map $h: Z \to \prod X_\alpha$ with $\pi_\alpha h = f_\alpha$ for all $\alpha \in A$.

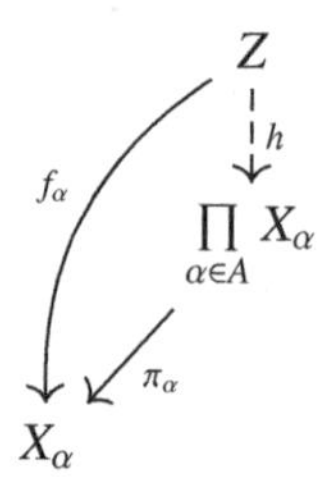

0.3.4 Products and Coproducts in Any Category

The universal properties outlined above provide a template for defining products and coproducts—and more generally limits and colimits—in any category. A more complete discussion is found in chapter 4, but for now, we'd like to point out that in an arbitrary category, products and coproducts may not exist, and when they do they might not look like disjoint unions or Cartesian products. For example, the category Fld of fields doesn't have products: if there were a field $\mathbf{k}$ that were the product of $\mathbb{F}_2$ and $\mathbb{F}_3$, then there would be homomorphisms $\mathbf{k} \to \mathbb{F}_2$ and $\mathbf{k} \to \mathbb{F}_3$. But this is impossible since the characteristic of $\mathbf{k}$ would be equal to both 2 and 3. On the other hand the category $\mathsf{Vect}_\mathbf{k}$ and more generally $R\mathsf{Mod}$ has both products and coproducts. Products are Cartesian products, but coproducts are direct sums. In Grp coproducts are free products, while in the category of abelian groups, coproducts are direct sums. Even in the category Set, there is something to say about the existence of products and coproducts. The *axiom of choice* is precisely the statement that for any nonempty collection of sets $\{X_\alpha\}_{\alpha\in A}$, the product $\prod X_\alpha$ exists and is nonempty. (For more on the axiom of choice, see section 3.4.)

Even though products and coproducts in an arbitrary category might look different than they do in Set, the constructions are closely related to the constructions in Set. That's because the universal properties of products and coproducts in an arbitrary category C yield bijections of sets

$$\mathsf{C}\left(\textstyle\coprod_\alpha X_\alpha, Z\right) \cong \textstyle\prod_\alpha \mathsf{C}(X_\alpha, Z) \quad\Big|\quad \mathsf{C}\left(Z, \textstyle\prod_\alpha X_\alpha\right) \cong \textstyle\prod_\alpha \mathsf{C}(Z, X_\alpha)$$

In other words, the product is characterized by the fact that maps *into* it are in bijection with maps into each of the factors. Dually, the coproduct is characterized by the fact that maps *out of* it are in bijection with maps out of each of the components. (In a sense that can be made precise, products and coproducts in a category are [co]representations of products and coproducts of sets.) To phrase it another way, coproducts come out of the first entry of hom as products, and products come out of the second entry of hom as products. An example where this comes up often is in $\mathsf{Vect}_\mathbf{k}$ and $R\mathsf{Mod}$ where the coproduct is direct sum and the product is Cartesian product. For an R module X, let $X^* := R\mathsf{Mod}(X, R)$ denote the dual space. Then setting $Z = R$ in the first isomorphism above yields

$$\left(\bigoplus X_\alpha\right)^* \quad\cong\quad \prod (X_\alpha)^*$$

So the fact that the "*the dual of the sum is the product of the duals*" is a consequence of the existence of coproducts in $R\mathsf{Mod}$.

0.3.5 Exponentiation in Set

In the category of sets, the set of morphisms $\mathsf{Set}(X, Y)$ is also denoted Y^X. Moreover there is a natural *evaluation* map eval $X\times\colon Y^X \to Y$ defined by $\text{eval}(x, f) = fx$. The exponential notation is convenient for expressing various isomorphisms, such as

$$(X \times Y)^Z \cong X^Z \times Y^Z$$

which is a concise way to express the universal property of the Cartesian product: maps from a set Z into a product correspond to maps from Z into the factors. There is also the isomorphism

$$Y^{X \times Z} \cong (Y^X)^Z$$

which is an expression of the *product-hom adjunction*. We made brief mention of this adjunction in example 0.6 and will discuss it in greater detail in example 5.1. But here's a sneak preview. Fix a set X. Let L be the functor $X \times -$ and let R be the functor $\mathsf{Set}(X, -)$. In this notation, the isomorphism $Y^{X \times Z} \cong (Y^X)^Z$ becomes $\mathsf{Set}(LZ, Y) \cong \mathsf{Set}(Z, RY)$, which evokes the defining property of *adjoint* linear maps. (Hence the term "adjunction.")

0.3.6 Partially Ordered Sets

A partially ordered set or *poset* is a set $\mathcal{P}$ together with a relation $\leq$ on $\mathcal{P}$ that is reflexive, transitive, and antisymmetric. Reflexive means that for all $a \in \mathcal{P}$, $a \leq a$; transitive means that for all $a, b, c \in \mathcal{P}$, if $a \leq b$ and $b \leq c$, then $a \leq c$; antisymmetric means that for all $a, b \in P$, if $a \leq b$ and $b \leq a$, then $a = b$.

One can view a poset as a category whose objects are the elements of $\mathcal{P}$ by declaring there to be a morphism $a \to b$ if and only if $a \leq b$. Transitivity says not only can composition be defined, but it may only be defined in *one* way since there's at most one morphism between objects. Alternatively, one can define a poset to be a (small) category with the property that there's at most one morphism between objects.

We expect that you have encountered different definitions for some of the concepts in this section than what we have provided. In your previous work, an injective function, for example, might have been defined in terms of what it does to the elements of the domain. Here, instead, we've defined an injective function in terms of how it interacts with other functions. For another example, instead of defining what the disjoint $X \coprod Y$ *is* (which ultimately involves the Zermelo-Fraenkel axioms of union and extension), we've characterized it up to isomorphism by explaining how it interacts with other sets, bringing back to mind the Lawvere quote that opened the chapter.

Exercises

1. Suppose $\mathcal{S}$ is a collection of subsets of X whose union equals X. Prove there is a coarsest topology $\mathcal{T}$ containing $\mathcal{S}$ and that the collection of all finite intersections of sets in $\mathcal{S}$ is a basis for $\mathcal{T}$. In this situation, the collection $\mathcal{S}$ is called *a subbasis* for the topology $\mathcal{T}$.

2. Prove that a function $f\colon X \to Y$ between topological spaces is continuous if and only if $f^{-1}B$ is open for every B in a basis for the topology on Y.

3. Here are some examples and short exercises about morphisms.

a) Prove that left-invertible morphisms are monic and right-invertible morphisms are epic.

b) Give an example of an epimorphism which is not right invertible.

c) Prove that if a morphism is left invertible and right invertible, then it is invertible.

d) Give an example of a morphism in Top that is epic and monic but not an isomorphism.

e) In some category, give an example of two objects X and Y that are not isomorphic but which nonetheless have monomorphisms: $X \rightleftarrows Y$.

4. Discuss the initial object, the terminal object, products, and coproducts in the categories Grp and $\mathsf{Vect}_{\mathbf{k}}$.

5. Prove the other part of theorem 0.1. That is, prove that $f\colon X \to Y$ is an isomorphism in a category C if and only if $f_*\colon \mathsf{C}(Z, X) \to \mathsf{C}(Z, Y)$ is an isomorphism for every object Z.

6. Prove the Yoneda lemma in section 0.2.3. The key is to observe that $\mathsf{C}(X, X)$ has a special element, namely id_X. So, for any natural transformation $\eta\colon \mathsf{C}(-, X) \to F$, one obtains a special element $\eta\,\mathrm{id}_X \in FX$, which completely determines η.

1 Examples and Constructions

All of it was written by Sammy.... I wrote nothing.
—Henri Cartan (Jackson, 1999)

Introduction. Our goal in this chapter is to construct new topological spaces from given ones. We'll do so by focusing on four basic constructions: subspaces in section 1.2, quotients in section 1.3, products in section 1.4, and coproducts in section 1.5. To maintain a categorical perspective, the discussion of each construction will fit into the following template:

- **The classic definition**: an explicit construction of the topological space
- **The first characterization**: a description of the topology as either the coarsest or the finest topology for which maps into or out of the space are continuous, leading to a better definition
- **The second characterization**: a description of the topology in terms of a universal property as given in theorems 1.1, 1.2, 1.3, and 1.4

Before we construct topological spaces, it will be good to have some examples in mind. We'll begin then in section 1.1 with examples of topological spaces and continuous maps between them.

1.1 Examples and Terminology

Let's open with examples of spaces followed by examples of continuous functions.

1.1.1 Examples of Spaces

Example 1.1 Any set X may be endowed with the *cofinite* topology, where a set U is open if and only its complement $X \setminus U$ is finite (or if $U = \varnothing$). Similarly, any set may be equipped with the *cocountable* topology whose open sets are those whose complement is countable.

Example 1.2 The empty set $\varnothing$ and the one-point set $*$ are topological spaces in unique ways. For any space X, the unique functions $\varnothing \to X$ and $X \to *$ are continuous. The empty set is initial and the one-point set is terminal in Top, just as they are in Set.

Example 1.3 As we saw in section 0.1, $\mathbb{R}$ is a topological space with the usual metric topology, but it admits other topologies, too. For example, like any set, $\mathbb{R}$ has a cofinite topology and a cocountable topology. The set $\mathbb{R}$ also has a topology with basis of open sets given by intervals of the form $[a, b)$ for $a < b$. This is called the *lower limit topology* (or the *Sorgenfrey topology*, or the *uphill topology*, or the *half-open topology*). Unless specified otherwise, $\mathbb{R}$ will be given the metric topology.

Example 1.4 In general, for any totally ordered set X, the intervals $(a, b) = \{x \in X \mid a < x < b\}$, along with the intervals (a, ∞) and $(-\infty, b)$, define a topology called the *order topology*. The set $\mathbb{R}$ is totally ordered and the order topology on $\mathbb{R}$ coincides with the usual topology.

Example 1.5 Unless otherwise specified, the natural numbers $\mathbb{N}$ and the integers $\mathbb{Z}$ are given discrete topologies, but there are others. Notably, there is a topology on $\mathbb{Z}$ for which the sets

$$S(a, b) \quad = \quad \{an + b \mid n \in \mathbb{N}\}$$

for $a \in \mathbb{Z} \setminus \{0\}$ and $b \in \mathbb{Z}$, together with $\varnothing$, are open. Furstenberg used this topology in a delightful proof noting that there are infinitely many primes (see Mercer (2009)). It's not hard to check that the sets $S(a, b)$ are also closed in this topology, and since every integer except ± 1 has a prime factor, it follows that

$$\mathbb{Z} \setminus \{-1, +1\} \quad = \quad \bigcup_{p \text{ prime}} S(p, 0)$$

Since the left hand side is not closed (no nonempty finite set can be open) there must be infinitely many closed sets in the union on the right. Therefore there are infinitely many primes!

Example 1.6 Let R be a commutative ring with unit and let $\operatorname{spec} R$ denote the set of prime ideals of R. The *Zariski topology* on $\operatorname{spec} R$ is defined by declaring the closed sets to be the sets of the form $VE = \{p \in \operatorname{spec} R \mid E \subseteq p\}$, where E is any subset of R.

Example 1.7 A *norm* on a real (or complex) vector space V is a function $|| - ||\colon V \to \mathbb{R}$ (or $\mathbb{C}$), satisfying

- $||v|| \geq 0$ for all vectors v with equality if and only if $v = 0$
- $||v + w|| \leq ||v|| + ||w||$ for all vectors v, w
- $||\alpha v|| = |\alpha| ||v||$ for all scalars α and vectors v

Every normed vector space is a metric space and hence a topological space with metric defined by $d(x, y) = ||x - y||$. The standard metric on $\mathbb{R}^n$ comes from the norm defined by $||(x_1, \ldots, x_n)|| := \sqrt{\sum_{i=1}^n |x_i|^2}$. More generally, for any $p \geq 1$, the *p-norm* on $\mathbb{R}^n$ is defined by

$$||(x_1, \ldots, x_n)||_p \quad := \quad \left(\textstyle\sum_{i=1}^n |x_i|^p\right)^{\frac{1}{p}}$$

and the *sup norm* is defined by

$$||(x_1, \ldots, x_n)||_\infty \quad := \quad \sup\{|x_1|, \ldots, |x_n|\}$$

These norms define different metrics with different open balls, but for any of these norms on $\mathbb{R}^n$, the passage *norm* $\rightsquigarrow$ *metric* $\rightsquigarrow$ *topology* leads to the same topology. In fact, for any choice of norm on a *finite* dimensional vector space, the corresponding topological spaces are the same—not just homeomorphic but literally the same.

Example 1.8 We can generalize the previous example from $\mathbb{R}^n$ to $\mathbb{R}^{\mathbb{N}}$, the space of sequences in $\mathbb{R}$, if we avoid those sequences with divergent norm. The set l_p of sequences $\{x_n\}$ for which $\sum_{n=1}^{\infty} x_n^p$ is finite is a subspace of $\mathbb{R}^{\mathbb{N}}$ (see section 1.2), and l_p with

$$||\{x_i\}||_p \quad := \quad (\sum_{i=1}^{\infty} |x_i|^p)^{\frac{1}{p}}$$

is a normed vector space. It's difficult to compare the topological spaces l_p for different p since the underlying sets are different. For instance, $\{1/n\}$ is in l_2 but not in l_1. Even so, the spaces l_p are homeomorphic as topological spaces (Kadets, 1967). The the set l_∞ of bounded sequences with $||\{x_i\}|| := \sup |x_i|$ is also a normed vector space, but it is not homeomorphic to l_p for $p \neq \infty$; it is an exercise to prove it.

Note that set $\mathbb{R}^{\mathbb{N}}$ of sequences $(x_1, x_2, \ldots)$ in $\mathbb{R}$ can be viewed as a topological space when endowed with the product topology (defined in section 1.4) and the subset of pth power summable sequences can be given a subspace topology (defined in section 1.2). The resulting topology is quite different from the topology obtained from viewing l_p as a normed vector space.

1.1.2 Examples of Continuous Functions

With some examples of spaces in tow, let's now consider a few examples of continuous functions. The first one we'll look at is a vivid illustration in Top of the philosophy (introduced in chapter 0) that objects are determined by their relationships with other objects.

Example 1.9 The set $S = \{0, 1\}$ with the topology $\{\varnothing, \{1\}, S\}$ is sometimes called the *Sierpiński two-point space.* In this topology, for any open set $U \subseteq X$, the characteristic function $\chi_U : X \to S$ defined by

$$\chi_U(x) = \begin{cases} 1 & \text{if } x \in U \\ 0 & \text{if } x \notin U \end{cases}$$

is a continuous function. What's more, every continuous function $f : X \to S$ is of the form χ_U where $U = f^{-1}\{1\}$. Thus the open subsets of X are in one-to-one correspondence with continuous functions $X \to S$. In other words the set $\mathsf{Top}(X, S)$ is a copy of the topology of X.

Example 1.10 The previous example shows that the topology of a space X can be recovered from the set of maps $\mathsf{Top}(X, S)$, so you might wonder: *can the points be recovered, too?* But that's an easy "Yes!" Since a point $x \in X$ is the same as a map $* \to X$, the set of points of a space X is isomorphic to the set $\mathsf{Top}(*, X)$.

A practical impact of the philosophy referred to above is that a space X can be studied by looking at continuous functions either to or from a (usually simpler) space. For example, the fundamental group of X, which we'll cover in chapter 6, involves functions from the circle S^1 to X. And sequences in X, which are used to probe topological properties (as we'll see in chapter 3), are continuous functions from the discrete space $\mathbb{N}$ to X. On the other hand, maps *out of* X are also interesting. For instance, maps from X to the discrete space $\{0, 1\}$ detect connectedness. For another instance involving maps into X, homotopy classes of maps $* \to X$ reveal path components. We'll discuss both connectedness and path connectedness in section 2.1. And speaking of paths....

Example 1.11 A *path* in a space X is a continuous function $\gamma\colon [0, 1] \to X$. A *loop* in a space X is a continuous function $\gamma\colon [0, 1] \to X$ with $\gamma 0 = \gamma 1$.

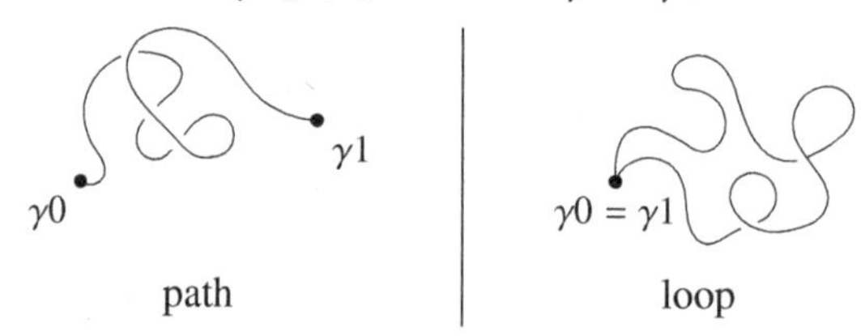

Example 1.12 If (X, d) is a metric space and $x \in X$, then the function $f\colon X \to \mathbb{R}$ defined by $fy = d(x, y)$ is continuous.

Example 1.13 Unlike the categories Grp and $\mathsf{Vect}_\mathbb{k}$ where bijective morphisms are isomorphisms, not every continuous bijection between topological spaces is a homeomorphism. For example, the identity function $\mathrm{id}\colon (\mathbb{R}, \mathcal{T}_{\text{discrete}}) \to (\mathbb{R}, \mathcal{T}_{\text{usual}})$ is a continuous bijection that is not a homeomorphism. But continuous bijections are always homeomorphisms in the category of compact Hausdorff spaces—see corollary 2.18.2.

Armed with examples of topological spaces and continuous functions, we now turn to the question of construction. How can we construct new spaces from existing ones? *Subspaces, quotients, products,* and *coproducts* are a few ways, and although the definitions of some (or perhaps all) of these constructions may be familiar, keep in mind that the goal of the remainder of the chapter is to view them through a categorical lens. As mentioned in the chapter introduction, we'll accomplish this by exploring each construction—a subspace, a quotient, a product, a coproduct—in three stages:

- **The classic definition**: the familiar, explicit construction of the topology
- **The first characterization**: a description of the topology as either the coarsest or the finest topology for which maps in to or out of the space are continuous, leading to a better definition.
- **The second characterization**: a description of the topology in terms of a universal property

As the text unwinds, take note of the words "finest" and "coarsest" in the first characterization. Which term appears in which of the four constructions? Also keep an eye out for which topologies are characterized by maps *into* the space and which are characterized by maps *out of* the space.

1.2 The Subspace Topology

Given a set X, we can obtain a new set by choosing a subset Y of X. If X is endowed with a topology, we'd like a way to see Y as a topological space, too. This leads to the first of the four constructions—a subspace. The subspace topology is often defined (for example, in Munkres (2000)) as follows:

Definition 1.1 Let $(X, \mathcal{T}_X)$ be a topological space and let Y be any subset of X. The *subspace topology* on Y is given by $\mathcal{T}_Y := \{U \cap Y \mid U \in \mathcal{T}_X\}$.

You can and are encouraged to check that this definition does indeed define a topology on Y. What's more interesting, though, is the property it satisfies. In particular, Y naturally comes with an inclusion map $i\colon Y \to X$, and the subspace topology is the coarsest topology on Y for which i is continuous. This is its first characterization.

1.2.1 The First Characterization

Before we elaborate on this characterization, let's consider a more general situation by way of motivation. Let $(X, \mathcal{T}_X)$ be a topological space and let S be any set whatsoever. Consider a function

$$f\colon S \to X$$

It makes no sense to ask if f is continuous until S is equipped with a topology. There always exist topologies on the set S that will make f continuous—the discrete topology is one. But is there a coarser one? Is there a *coarsest* one? The answer to both questions is "yes." Indeed, the intersection of any topologies on S for which f is continuous is again a topology on S for which f is continuous. Therefore, the intersection of *all* topologies on S for which f is continuous will be the coarsest topology for which f is continuous. Let's call it $\mathcal{T}_f$ and observe that it has the simple description $\{f^{-1}U \mid U \subseteq X \text{ is open}\}$. This shows that the subspace topology $\mathcal{T}_Y$ on a subset $Y \subseteq X$ is the same as $\mathcal{T}_i$ where $i\colon Y \to X$ is the natural inclusion. This prompts us to adopt a better definition for the subspace topology.

Better definition Let $(X, \mathcal{T}_X)$ be a topological space and let Y be any subset of X. The *subspace topology* on Y is the coarsest topology on Y for which the canonical inclusion $i\colon Y \hookrightarrow X$ is continuous.

More generally, if S is any set and if $f\colon S \to X$ is an injective function, then $\mathcal{T}_f$—the coarsest topology on S for which f is continuous—may be called *the subspace topology* on S. This is a good definition even though the set S is not necessarily a subset of X.

Why? Since f is injective, S is isomorphic *as a set* to its image $fS \subseteq X$, and the space $(S, \mathcal{T}_f)$ determined by $f\colon S \to X$ is homeomorphic to $fS \subseteq X$ with the subspace topology determined by the inclusion $i\colon fS \hookrightarrow X$. (If $f\colon S \to X$ is not injective, then there is still a coarsest topology $\mathcal{T}_f$ on S that makes f continuous, though we won't refer to it as the *subspace* topology.)

Definition 1.2 Suppose $f\colon Y \to X$ is a continuous injection between topological spaces. The map f is called an *embedding* when the topology on Y is the same as the subspace topology $\mathcal{T}_f$ induced by f.

Example 1.14 Consider the set $[0,1]$ with the discrete topology. The evident map $i\colon ([0,1], \mathcal{T}_{\text{discrete}}) \to (\mathbb{R}, \mathcal{T}_{\text{ordinary}})$ is a continuous injection, but it is not an embedding. The topology on the domain is not the subspace topology induced by i.

Notice that endowing a subset $Y \subseteq X$ with the subspace topology provides meaning to the question, "Is a function $Z \to Y$ continuous?" Simply put, the subspace topology determines continuous maps to Y. The converse holds as well: continuous maps to Y determine the topology on Y. This is yet another illustration of the philosophy that objects in a category are determined by morphisms to and from them. It's also the heart of the second characterization of the subspace topology.

1.2.2 The Second Characterization

This way of thinking about the subspace topology describes the important universal property which characterizes precisely which functions into the subspace are continuous—they are, reasonably, the functions $Z \to Y$ that are continuous when regarded as functions into X.

Theorem 1.1 Let $(X, \mathcal{T}_X)$ be a topological space, let Y be a subset of X, and let $i\colon Y \hookrightarrow X$ be the natural inclusion. The subspace topology on Y is characterized by the following property:

Universal property for the subspace topology For every topological space $(Z, \mathcal{T}_Z)$ and every function $f\colon Z \to Y$, f is continuous if and only if the map $if\colon Z \to X$ is continuous.

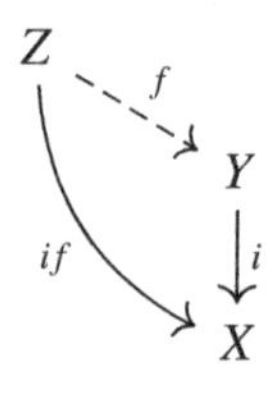

Proof. Let's think of this theorem in two parts: we'll first verify that the subspace topology has the universal property. Then we'll verify that the subspace topology is *characterized by* this universal property, which is to say that any topology on Y satisfying the universal property must be the subspace topology.

To start, let $\mathcal{T}_Y$ be the subspace topology on Y, let $(Z, \mathcal{T}_Z)$ be any topological space, and let $f: Z \to Y$ be a function. We have to prove that $f: Z \to Y$ is continuous if and only if $if: Z \to X$ is continuous. First, if f is continuous, then the composition of continuous functions $if: Z \to X$ is also continuous. Now suppose $if: Z \to X$ is continuous, and let U be any open set in Y. Then $U = i^{-1}V$ for some open $V \subseteq X$. Since if is continuous, the set $(if)^{-1}V \subseteq Z$ is open in Z. And since $(if)^{-1}V = f^{-1}U$, it follows that $f^{-1}U$ is open and so $f: Z \to Y$ is continuous. The topology $\mathcal{T}_Y$ therefore has the universal property above.

Suppose now that $\mathcal{T}'$ is any topology on Y having the universal property. We'll prove that $\mathcal{T}'$ equals the subspace topology $\mathcal{T}_Y$, that is $\mathcal{T}' \subseteq T_Y$ and $\mathcal{T}_Y \subseteq \mathcal{T}'$. The universal property for $\mathcal{T}'$ is that for every topological space $(Z, \mathcal{T}_Z)$ and for any function $f: Z \to Y$, the map f is continuous if and only if if is continuous. In particular, if we let $(Z, \mathcal{T}_Z)$ be $(Y, \mathcal{T}_Y)$ where $\mathcal{T}_Y$ is the subspace topology on Y and let $f: Y \to Y$ be the identity function, then we have the following diagram

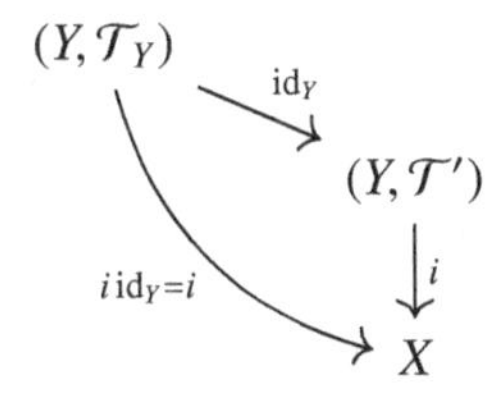

Since we know the function $i\,\mathrm{id}_Y = i: Y \to X$ is continuous when Y has the subspace topology $\mathcal{T}$, the universal property implies that $\mathrm{id}_Y: (Y, \mathcal{T}_Y) \to (Y, \mathcal{T}')$ is continuous, and therefore the subspace topology $\mathcal{T}_Y$ is finer than $\mathcal{T}'$, that is $\mathcal{T}' \subseteq \mathcal{T}_Y$. Finally, to show that $\mathcal{T}_Y \subseteq \mathcal{T}'$, let $(Z, \mathcal{T}_Z)$ be $(Y, \mathcal{T}')$ and let $f = \mathrm{id}_Y: (Y, \mathcal{T}') \to (Y, \mathcal{T}')$. So we have the following diagram

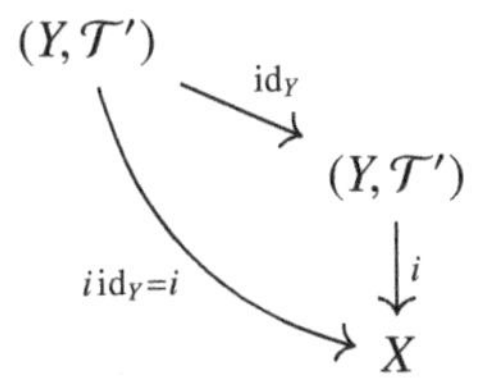

The continuity of id_Y implies that $i\,\mathrm{id}_Y = i: Y \to X$ is also continuous, and so $\mathcal{T}'$ is a topology on Y for which the inclusion $i: Y \to X$ is continuous. But the subspace topology $\mathcal{T}_Y$ is the coarsest topology on Y for which $i: Y \to X$ is continuous, and therefore $\mathcal{T}_Y$ is coarser than $\mathcal{T}'$. In other words $\mathcal{T}_Y \subseteq \mathcal{T}'$. □

Example 1.15 In the subspace topology on $\mathbb{Q} \subseteq \mathbb{R}$, open sets are of the form $\mathbb{Q} \cap (a, b)$ whenever $a < b$. Notice that the discrete and subspace topologies on $\mathbb{Q}$ are not equivalent: for any rational r, the singleton set $\{r\}$ is open in the former but not in the latter.

1.3 The Quotient Topology

Before getting to the definition and characterizations of the quotient topology, let's recall how quotients of sets work. Suppose X is a set and let $\sim$ be an equivalence relation on X. Then $X/\sim$ denotes the set of equivalence classes, and the natural projection $\pi\colon X \twoheadrightarrow X/\sim$ that sends x to its equivalence class defines a surjective function whose fibers are the equivalence classes of $\sim$.

Conversely, given any set S and any surjection $\pi\colon X \twoheadrightarrow S$, the set S is isomorphic to $X/\sim$ where $\sim$ is the equivalence relation whose equivalence classes are the fibers of π:

$$x \sim y \iff \pi x = \pi y$$

The map π conveniently provides the isomorphism

$$\begin{aligned} S &\xrightarrow{\cong} X/\sim \\ s &\longmapsto \pi^{-1}s \end{aligned}$$

Now suppose X is a topological space. So we have a surjective map $\pi\colon X \twoheadrightarrow S$ from a topological space X to a set S. What kind of topology can or should we give the set S? This topology—called the quotient topology—is often defined as follows.

Definition 1.3 A set $U \subseteq S$ is open in the *quotient topology* if and only if $\pi^{-1}U$ is open in X.

Because S and $X/\sim$ (the quotient determined by the fibers of π) aren't even distinguishable as sets, we can always think of the quotient topology as being defined either on S *or* on $X/\sim$. This is analogous to thinking of the subspace topology determined by an injection $f\colon S \hookrightarrow X$ as either being defined on S *or* as being defined on the subset $fS \subseteq X$.

1.3.1 The First Characterization

It doesn't make sense to ask if π is continuous when $\pi\colon X \twoheadrightarrow S$ is a map from a topological space X to a set S. We can ask whether there exists a topology on S for which $\pi : X \twoheadrightarrow S$ continuous. The answer to this question is a straightforward "yes" if S is endowed with the indiscrete topology. But is there a finer topology? Is there a *finest* topology? Again, the answer is affirmative. In fact, definition 1.3 makes the quotient topology on S the finest topology for which the map $\pi\colon X \to S$ is continuous: declaring U to be open *only if* $\pi^{-1}U$ is open implies that π is continuous, while declaring U to be open if $\pi^{-1}U$ is open makes the quotient topology the finest topology for which π is continuous. We thus obtain the first characterization of the quotient topology, which is a better definition.

Better definition Let X be a topological space, let S be a set, and let $\pi\colon X \twoheadrightarrow S$ be surjective. The *quotient topology* on S is the finest topology for which π is continuous, and π is called a *quotient map*.

A word of caution: be careful when talking about the finest topology satisfying some property since such a topology may not exist. This is less of an issue for the *coarsest* topology satisfying a property. The difference is that the intersection of topologies is always a topology, whereas the union of topologies is usually not a topology.

1.3.2 The Second Characterization

In section 1.2.2, we observed that a topology on a subset $Y \subseteq X$ is determined by specifying what $\mathsf{Top}(Z, Y)$ is for any space Z. Analogously, given a surjection $\pi\colon X \twoheadrightarrow S$ from a space to a set, a topology on S is determined by specifying what $\mathsf{Top}(S, Z)$ is for any space Z. The universal property that characterizes the quotient topology on S tells us that the continuous maps $S \to Z$ are precisely those whose precompositon with π are continuous functions $X \to Z$.

Theorem 1.2 Let X be a topological space, let S be a set, and let $\pi\colon X \to S$ be surjective. The quotient topology on S is determined by the following property.

Universal property for the quotient topology For every topological space Z and every function $f\colon S \to Z$, f is continuous if and only if $f\pi\colon X \to Z$ is continuous.

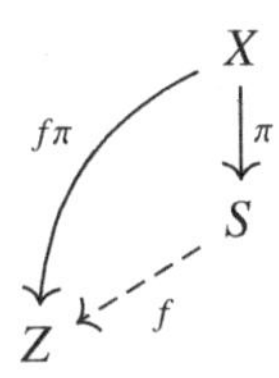

Proof. Exercise. □

The universal property of the quotient topology tells us precisely which functions $S \to Z$ from a quotient to a space Z are continuous: they are continuous maps $X \to Z$ that are constant on the fibers of $\pi\colon X \to S$.

Example 1.16 The map $\pi\colon [0, 1] \to S^1$ defined by $\pi(t) = (\cos(2\pi t), \sin(2\pi t))$ is a quotient map. Therefore, for any space Z, continuous functions $S^1 \to Z$ are the same as continuous functions $[0, 1] \to Z$ which factor through π. That is, continuous functions $S^1 \to Z$ are the same as paths $\gamma\colon [0, 1] \to Z$ satisfying $\gamma 0 = \gamma 1$. These are the loops in Z.

Example 1.17 The *projective space* $\mathbb{RP}^n$ is defined to be the quotient of $\mathbb{R}^{n+1} \smallsetminus \{0\}$ by the relation $x \sim \lambda x$ for $\lambda \in \mathbb{R}$. So $\mathbb{RP}^n$ is the set of all lines through the origin in $\mathbb{R}^{n+1}$, and the quotient topology gives us the topology on this set of lines.

Example 1.18 As in the previous example, topological spaces are often constructed by starting with a familiar space and then identifying points to obtain a quotient. In the figure below, for instance, a new space is obtained from the unit square I^2 in $\mathbb{R}^2$ by identifying opposite sides. The topology on $I^2/\!\sim$ is obtained from the mapping $I^2 \to I^2/\!\sim$ where

$(x, 0) \sim (x, 1)$ and $(0, y) \sim (1, y)$.

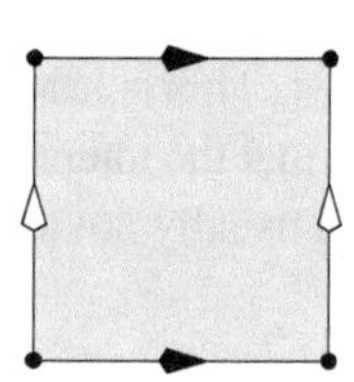

The resulting quotient space is called the *torus*, T. Other identifications of the square yield the *Mobius band* M, the *Klein bottle* K, and the *projective plane* $\mathbb{RP}^2$:

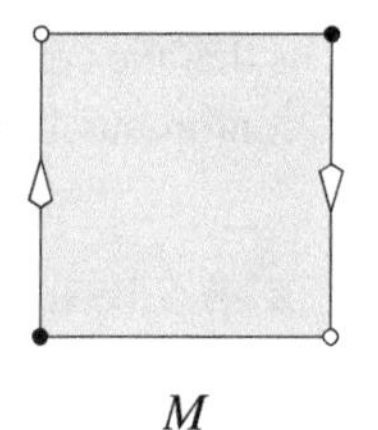

M

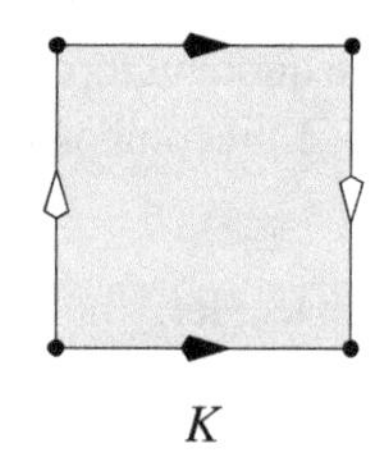

K

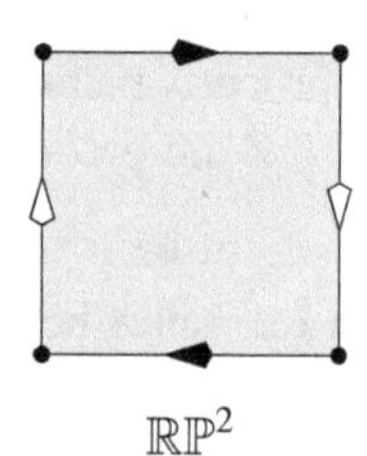

$\mathbb{RP}^2$

Notice now that we have two definitions of $\mathbb{RP}^2$: example 1.17 describes it as the space of lines through the origin, and here we've described it as a quotient of the unit square. We encourage you to verify that these two descriptions yield homeomorphic spaces.

1.4 The Product Topology

Let $\{X_\alpha\}_{\alpha\in A}$ be an arbitrary collection of topological spaces and consider the set

$$X = \prod_{\alpha\in A} X_\alpha$$

We'd like to make the set X into a topological space, but how? One way is by equipping it with the product topology, typically defined as follows.

Definition 1.4 The *product topology* on X is defined to be the topology generated by the basis

$$\left\{ \prod_{\alpha\in A} U_\alpha \;\middle|\; U_\alpha \subseteq X_\alpha \text{ is open and all but finitely many } U_\alpha = X_\alpha \right\}$$

But this definition, with its surprising "all but finitely many," suggests that there are better ways to define the product topology. And indeed, there are.

1.4.1 The First Characterization

Recall from section 0.3.3 that the set X comes with projection maps $\pi_\alpha : X \to X_\alpha$. Is there a topology on X for which these natural maps are continuous? The discrete topology is certainly one, and the intersection of all topologies that make the projections continuous will be the coarsest topology for which the projections are continuous.

Better definition Let $\{X_\alpha\}_{\alpha\in A}$ be an arbitrary collection of topological spaces and let $X = \prod_{\alpha\in A} X_\alpha$. The *product topology on* X is defined to be the coarsest topology on X for which all of the projections π_α are continuous.

The better definition of the product topology is equivalent to definition 1.4, and we'll leave the proof as an exercise.

1.4.2 The Second Characterization

The second characterization of the product topology amounts to saying precisely which functions to the product are continuous. As before, let $\{X_\alpha\}_{\alpha\in A}$ be an arbitrary collection of topological spaces, and consider the set $X = \prod_{\alpha\in A} X_\alpha$. Keeping in mind that the universal property of the product of sets says that functions into X are the same as collections of functions into the sets X_α, it's not hard to guess that for any space Z, a map $Z \to X$ is continuous whenever all the components $Z \to X \to X_\alpha$ are continuous.

Theorem 1.3 Let $\{X_\alpha\}_{\alpha\in A}$ be an arbitrary collection of topological spaces, and let $X = \prod_{\alpha\in A} X_\alpha$. Let $\pi_\alpha\colon X \to X_\alpha$ denote the natural projection. The product topology on X is characterized by the following property.

Universal property for the product topology For every topological space Z and every function $f\colon Z \to X$, f is continuous if and only if for every $\alpha \in A$, the component $\pi_\alpha f\colon Z \to X_\alpha$ is continuous.

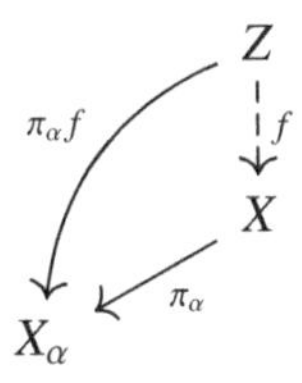

Proof. Exercise. □

Example 1.19 Let $X = \mathbb{R}^2$. One can write any function $f\colon S \to \mathbb{R}^2$ in terms of component functions $fs = (xs, ys)$, where the components xs and ys are simply given by the composition

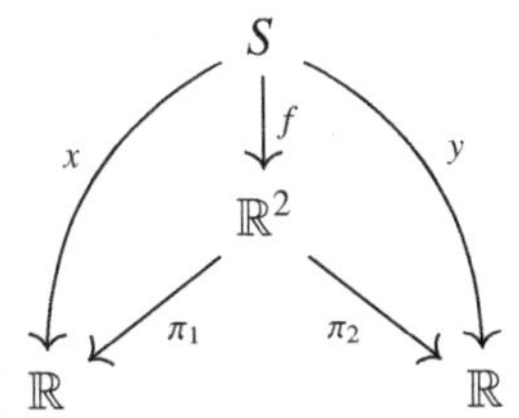

The function f is continuous if and only if x and y are continuous, and it's good to realize that this way of specifying which functions into $\mathbb{R}^2$ are continuous completely determines the topology on $\mathbb{R}^2$.

But be careful: functions from $\mathbb{R}^2$ and more generally $\mathbb{R}^n$ can be confusing, in part because our familiarity with $\mathbb{R}^n$ can give unjustified topological importance to the maps $\mathbb{R} \to \mathbb{R}^2$ given by fixing one of the coordinates. So don't make the mistake of thinking that a function $f: \mathbb{R}^2 \to S$ is continuous if the maps $x \mapsto f(x, y_0)$ and $y \mapsto f(x_0, y)$ are continuous for every x_0 and y_0, as in the diagram below:

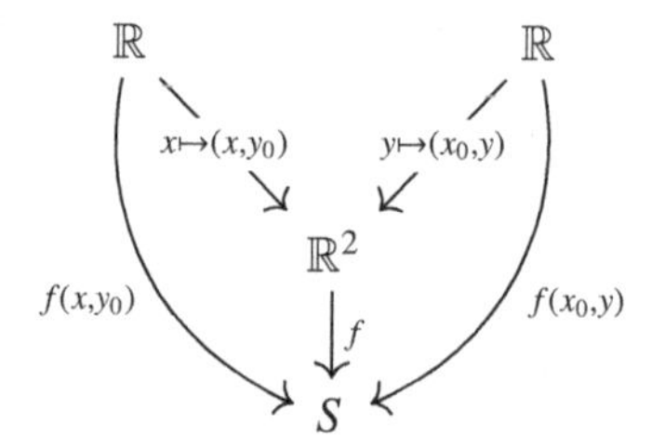

Here's a counterexample to keep in mind: the function $f: \mathbb{R}^2 \to \mathbb{R}$ defined by

$$f(x,y) = \begin{cases} \dfrac{xy}{x^2+y^2} & \text{if } (x,y) \neq (0,0) \\ 0 & \text{if } (x,y) = (0,0) \end{cases}$$

is not continuous even though for any choice of x_0 or y_0, the maps $f(x, y_0)$ and $f(x_0, y)$ are continuous functions $\mathbb{R} \to \mathbb{R}$.

1.5 The Coproduct Topology

Let $\{X_\alpha\}_{\alpha \in A}$ be a collection of topological spaces. We'd like to make the disjoint union $X = \coprod_{\alpha \in A} X_\alpha$ into a topological space. Typically, this is done via the following explicit definition of the coproduct topology.

Definition 1.5 A set $U \subseteq X$ is open in the *coproduct topology* if and only if it is of the form $U = \coprod_{\alpha \in A} U_\alpha$ where each $U_\alpha \subseteq X_\alpha$ is open.

But since the disjoint union—when viewed as a set—comes with canonical inclusion functions $i_\alpha: X_\alpha \to \coprod X_\alpha$ for each α, we'd like a topology on X for which these natural maps are continuous.

1.5.1 The First Characterization

There are many topologies on $\coprod X_\alpha$ for which the inclusions i_α are continuous—the indiscrete topology is one. But the right topology to put on $\coprod X_\alpha$ is the finest topology for which the maps $X_\alpha \to \coprod X_\alpha$ are all continuous. This leads to another definition—one that is equivalent to definition 1.5 but better.

Better definition Let $\{X_\alpha\}_{\alpha \in A}$ be an arbitrary collection of topological spaces, and let $X = \coprod_{\alpha \in A} X_\alpha$. The *coproduct topology on* X is defined to be the finest topology on X for which all of the inclusions i_α are continuous.

1.5.2 The Second Characterization

To characterize the coproduct topology a second way, recall from section 0.3.3 that functions from X, viewed as a set, are determined by collections of functions from X_α. That is, for any set Z, a collection of functions $f_\alpha \colon X_\alpha \to Z$ corresponds uniquely to a function $X \to Z$. Not surprisingly, then, the coproduct topology on X is characterized by the following universal property.

Theorem 1.4 Let $\{X_\alpha\}_{\alpha \in A}$ be an arbitrary collection of topological spaces and let $X = \coprod_{\alpha \in A}$. Let $i_\alpha \colon X_\alpha \to X$ denote the natural inclusion. The coproduct topology on X is characterized by the following property.

Universal property for the coproduct topology For every topological space Z and every function $f \colon X \to Z$, f is continuous if and only if for every $\alpha \in A$, $fi_\alpha \colon X_\alpha \to Z$ is continuous.

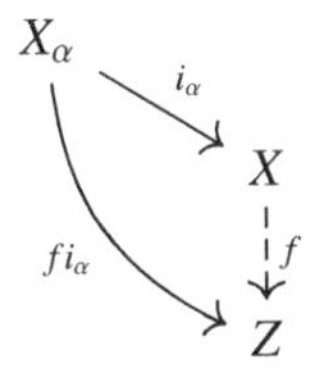

Proof. Exercise. □

Example 1.20 Any set X is the coproduct over its points viewed as singletons:

$$X \cong \coprod_{x \in X} \{x\}$$

As topological spaces, however, X is homeomorphic to $\coprod_{x \in X}\{x\}$ if and only if X has the discrete topology.

Here's a summary of the results we've collected thus far:

- The *subspace topology* on a subset $Y \subseteq X$ is the *coarsest* topology for which the natural inclusion $Y \hookrightarrow X$ is continuous. It's determined by maps *into* the subspace.
- The *quotient topology* on a quotient $X/\!\sim$ is the *finest* topology for which the natural projection $X \twoheadrightarrow X/\!\sim$ is continuous. It's determined by maps *out of* the quotient.
- The *product topology* on a set $\prod_{\alpha \in A} X_\alpha$ is the *coarsest* topology for which the natural projections $\prod_{\alpha \in A} X_\alpha \to X_\alpha$ are continuous. It's determined by maps *into* the product.

- The *coproduct topology* on a set $\coprod_{\alpha \in A} X_\alpha$ is the *finest* topology for which the natural inclusions $X_\alpha \to \coprod_{\alpha \in A} X_\alpha$ are continuous. It's determined by maps *out of* the coproduct.

You'll notice that "coarsest" and "into" are paired in the first and third constructions while "finest" and "out of" are paired in the second and fourth. This duality is no coincidence. Each of these four constructions is a special case of a more general, categorical construction—either a *limit* or a *colimit*. Limits are characterized by maps into them; colimits are characterized by maps out of them. We'll discuss these constructions in much greater detail in chapter 4, but exercises 1.12 and 1.13 at the end of the chapter provide a sneak peek.

1.6 Homotopy and the Homotopy Category

We'll close this chapter with an important application of the product topology, namely homotopy. A *homotopy* from a map $f: X \to Y$ to a map $g: X \to Y$ is a continuous function $h: X \times [0,1] \to Y$ satisfying $h(x,0) = fx$ and $h(x,1) = gx$. (Notice that "the topological space $X \times [0,1]$" makes sense now that we have the product topology!) Two maps $f, g: X \to Y$ are said to be *homotopic* if there is a homotopy between them, in which case we write $f \simeq g$. Homotopy defines an equivalence relation on the maps $\mathsf{Top}(X,Y)$, the equivalence classes of which are called *homotopy classes* of maps. We use $[f]$ to denote the homotopy class of f, and we use the notation $[X,Y]$ for the set of homotopy classes of maps. You can check that the composition of homotopic maps are homotopic, and thus composition of homotopy classes of maps is well defined. The *homotopy category* of topological spaces is defined to be the category hTop whose objects are topological spaces and whose morphisms are homotopy classes of maps:

$$\mathsf{hTop}(X,Y) := [X,Y]$$

Two spaces are said to be *homotopic* if and only if they are isomorphic in hTop. That is, X and Y are homotopic if and only if there exist maps $f: X \to Y$ and $g: Y \to X$ so that $gf \simeq \mathrm{id}_X$ and $fg \simeq \mathrm{id}_Y$. In this case we write $X \simeq Y$. A *homotopy invariant* is a property that is invariant under homotopy equivalence. More precisely, there is a natural functor

$$\mathsf{Top} \to \mathsf{hTop}$$

with $X \mapsto X$ and $f \mapsto [f]$. Functors from hTop are homotopy invariants, and functors from the category Top that factor through the functor $\mathsf{Top} \to \mathsf{hTop}$ are called *homotopy functors*.

Example 1.21 The space $\mathbb{R}^n$ is homotopic to the one-point space, $\mathbb{R}^n \simeq *$. To see this, define $f: * \to \mathbb{R}^n$ by $* \mapsto 0$ and define $g: \mathbb{R}^n \to *$ in the only way possible. Then $gf = \mathrm{id}_*$, and $fg: \mathbb{R}^n \to 0$ is homotopic to $\mathrm{id}_{\mathbb{R}^n}$ via the homotopy $h: \mathbb{R}^n \times [0,1] \to \mathbb{R}^n$

given by $h(x, t) = tx$. Spaces such as $\mathbb{R}^n$ that are homotopic to a point are said to be *contractible*.

Quite often, a restricted notion of homotopy applies. For example, if $\alpha, \beta\colon [0,1] \to X$ are paths from x to y, then a *homotopy of paths* is defined to be $h\colon [0,1]\times[0,1] \to Y$ satisfying $h(t,0) = \alpha t$, $h(t,1) = \beta t$, and $h(0,s) = x$ and $h(1,s) = y$ for all $s, t \in [0,1]$. In other words, the homotopy fixes the endpoints of the path: for all s, the path $t \mapsto h(t,s)$ is a path from x to y agreeing with α at $s = 0$ and with β at $s = 1$. At this point, you might wonder why we're interested in fixing the endpoints of paths throughout a homotopy of paths. The reason is simple. Without this requirement, way too many paths may become homotopic: you could continuously "wind in" the endpoint of a path until it meets the initial point, then move the initial point around, then expand the point to be another path.

Succinctly put, homotopy in topology is of great importance. We'll return to it in much more detail in chapter 6.

Exercises

1. Draw a diagram of all the topologies on a three-point set indicating which are contained in which.

2. In this chapter, $\mathbb{R}^n$ has been considered a topological space in two ways: as a metric space with the usual distance function and as the product of n copies of $\mathbb{R}$. Prove that these are the same.

3. Check that the Zariski topology does in fact define a topology on spec R, and sketch a picture of spec $\mathbb{C}[x]$ and spec $\mathbb{Z}$. For a more challenging problem, sketch a good picture of $\mathbb{Z}[x]$.

4. Give an example of a path $p\colon [0,1] \to X$ connecting a to b in the space $(X, \mathcal{T})$, where:

$$X = \{a, b, c, d\} \quad \text{and} \quad \mathcal{T} = \Big\{\varnothing, \{a\}, \{c\}, \{a, c\}, \{a, b, c\}, \{a, d, c\}, X\Big\}$$

5. Prove that any two norms on a finite dimensional vectors space (over $\mathbb{R}$ or $\mathbb{C}$) give rise to homeomorphic topological spaces.

6. Prove that l_∞ is not homeomorphic to l_p for $p \neq \infty$.

7. Let $C([0,1])$ denote the set of continuous functions on $[0,1]$. The following define norms on $C([0,1])$:

$$\begin{aligned} \|f\|_\infty &= \sup_{x\in[0,1]} |f(x)| \\ \|f\|_1 &= \int_0^1 |f| \end{aligned}$$

Prove that the topologies on $C([0,1])$ coming from these two norms are different.

8. Prove theorem 1.3. That is, prove that $X := \prod_{\alpha\in A} X_\alpha$ with the product topology has the universal property. Then prove that if X is equipped with any topology having the universal property, then that topology must be the product topology.

9. Are the subspace and product topologies consistent with each other?

Let $\{X_\alpha\}_{\alpha\in A}$ be a collection of topological spaces, and let $\{Y_\alpha\}$ be a collection of subsets; each $Y_\alpha \subseteq X_\alpha$. There are two things you can do to put a topology on $Y := \prod_{\alpha\in A} Y_\alpha$:

a) You can take the subspace topology on each Y_α, then form the product topology on Y.

b) You can take the product topology on X, view Y as a subset of X, and equip it with the subspace topology.

Is the outcome the same either way? If yes, prove it using only the universal properties. If no, give a counterexample.

10. Prove that the quotient topology is characterized by the universal property given in section 1.3.

11. Are the quotient and product topologies compatible with each other?

Let $\{X_\alpha\}_{\alpha\in A}$ be a collection of topological spaces, let $\{Y_\alpha\}_{\alpha\in A}$ be a collection of sets, and let $\{\pi_\alpha\colon X_\alpha \twoheadrightarrow Y_\alpha\}_{\alpha\in A}$ be a collection of surjections. Let $X = \prod_\alpha X_\alpha$; notice that you have a surjection $\pi\colon X \twoheadrightarrow Y$. There are two ways to put a topology on $Y := \prod_{\alpha\in A} Y_\alpha$:

a) Take the quotient topology on each Y_α, then form the product topology on Y.

b) Take the product topology on X, then put the quotient topology on Y.

Is the outcome the same either way? If yes, prove it using only the universal properties. If no, give a counterexample.

12. Suppose X is a topological space and $f\colon X \twoheadrightarrow S$ is surjective. Define an equivalence relation on X by $x \sim x' \iff fx = fx'$. Let

$$R = \{(x, x') \in X \times X \mid fx = fx'\}$$

There are two maps, call them $r_1\colon R \to X$ and $r_2\colon R \to X$, defined by the composition of inclusion $R \hookrightarrow X \times X$ with the two natural projections $X \times X \to X$.

$$R \hookrightarrow X \times X \underset{\pi_2}{\overset{\pi_1}{\rightrightarrows}} X$$

Learn what a coequalizer is and prove that the set S with the quotient topology is the coequalizer of r_1 and r_2.

13. Suppose X is a topological space and $f\colon S \hookrightarrow X$ is injective. Let $X/\!\sim$ be the quotient space generated by the equivalence relation $x \sim y \iff x, y \in fS$. Then there is a diagram:

$$X \underset{c}{\overset{\pi}{\rightrightarrows}} X/\!\sim$$

where π is the natural projection sending an element to its equivalence class and c is the constant map sending each $x \in X$ to the equivalence class of fS. Learn what an equalizer is and prove that the set S with the subspace topology is the equalizer of π and c.

14. This exercise starts off with a definition:

Definition 1.6 Let X and Y be topological spaces. A function $f\colon X \to Y$ is called *open* (or *closed*) if and only if fU is open (or closed) in Y whenever U is open (or closed) in X.

Let $(X, \mathcal{T}_X)$ and $(Y, \mathcal{T}_Y)$ be topological spaces, and suppose $f\colon X \to Y$ is a continuous surjection.

a) Give an example to show that f may be open but not closed.

b) Give an example to show that f may be closed but not open.

c) Prove that if f is either open or closed, then that the topology $\mathcal{T}_Y$ on Y is equal to $\mathcal{T}_f$, the quotient topology on Y.

15. Consider the closed disc D^2 and the two-sphere S^2:

$$\begin{aligned} D^2 &= \{(x, y) \in \mathbb{R}^2 \mid x^2 + y^2 \leq 1\} \\ S^2 &= \{(x, y, z) \in \mathbb{R}^3 \mid x^2 + y^2 + z^2 = 1\} \end{aligned}$$

Consider the equivalence relation on D^2 defined by identifying every point on $S^1 \subseteq D^2$. So each point in $D^2 \setminus S^1$ is a one point equivalence class, and the entire boundary ∂D^2 is one equivalence class. Prove that the quotient $D^2/\!\sim$ with the quotient topology is homeomorphic to S^2.

2 Connectedness and Compactness

One of the classical aims of topology is to classify topological spaces by their topological type, or in other terms to find a complete set of topological invariants.
—Samuel Eilenberg (1949)

Introduction. In chapter 1, we discussed four main constructions of topological spaces: subspaces, quotients, products, and coproducts. In this chapter, we'll see how these constructions interact with three main topological properties: connectedness, Hausdorff, and compactness. That is, are subspaces of compact spaces also compact? Is the quotient of a Hausdorff space itself Hausdorff? Are products of connected spaces also connected? Is a union of connected spaces connected? We'll explore these questions and more in the pages to come.

Section 2.1 contains a survey of basic notions, theorems, and examples of connectedness. It also includes a statement and categorical proof of the one-dimensional version of Brouwer's well-known fixed-point theorem. Section 2.2 contains the Hausdorff property, though we'll keep the discussion brief. The Hausdorff property becomes much richer once it's combined with compactness, which is the content of section 2.3. The same section also introduces three familiar theorems—the Bolzano-Weierstrass theorem, the Heine-Borel theorem, and Tychonoff's theorem.

2.1 Connectedness

We'll begin with a discussion of the main ideas about connectedness. The definitions are collected up front and the main results follow. The proofs are mostly left as exercises, but they can be found in most any classic text on topology, such as Willard (1970), Munkres (2000), Kelley (1955), Lipschutz (1965).

2.1.1 Definitions, Theorems, and Examples

Definition 2.1 A topological space X is *connected* if and only if one of the following equivalent conditions holds:

(i) X cannot be expressed as the union of two disjoint nonempty open sets.

(ii) Every continuous function $f\colon X \to \{0, 1\}$ is constant, where $\{0, 1\}$ is equipped with the discrete topology.

Exercise 2.1 at the end of the chapter asks you to prove the equivalence of the two definitions. Even though they are equivalent, we prefer the second. We can define an equivalence relation $\sim'$ on X by declaring $x \sim' y$ if and only if there's a connected subspace of X that contains both x and y. Reflexivity and symmetry are immediate, while transitivity follows from theorem 2.3. The equivalence classes of $\sim'$ are called the *connected components* of X. But there is also a different—and richer—kind of connectedness.

Definition 2.2 A topological space X is said to be *path connected* if and only if for all $x, y \in X$ there is a *path* that connects x and y.

Recall that a path from x to y in a topological space X is a map $\gamma\colon I \to X$ with $\gamma 0 = x$ and $\gamma 1 = y$. There is an equivalence relation on X defined by declaring $x \sim y$ if and only if there is a path in X connecting x and y. The existence of the constant path shows $\sim$ is reflexive. To see that it is symmetric, suppose f is a path from x to y. Then g defined by $gt = f(1-t)$ is a path from y to x. For transitivity, we first define the product of paths. If f is a path from x to y and g is a path from y to z, the *product* $g \cdot f$ is the path from x to z obtained by first traversing f from x to y and then traversing g from y to z, each at twice the speed:

$$(g \cdot f)t = \begin{cases} f(2t) & 0 \leq t \leq \frac{1}{2} \\ g(2t-1) & \frac{1}{2} \leq t \leq 1 \end{cases} \tag{2.1}$$

This shows that $\sim$ is transitive, and the equivalence classes of $\sim$ are called the *path components* of X. In essence, path components are homotopy classes of maps $* \to X$ since a point $x \in X$ is a map $* \to X$ and a path between two points $* \to X$ is a homotopy between the maps. We will denote the set of all path components in X by $\pi_0 X$.

Equipped with basic definitions, we now list some of the theorems. Commentary will be kept to a minimum as this section is meant to be a highlight of standard results. Do, however, take special notice of our frequent use of condition (ii) in lieu of condition (i) from definition 2.1.

Theorem 2.1 If X is (path) connected and $f\colon X \to Y$, then fX is (path) connected.

Proof. If fX is not connected, then there is a nonconstant map $g\colon fX \to \{0, 1\}$, which implies the map $gf\colon X \to \{0, 1\}$ is not constant. Now suppose X is path connected. Let $y, y' \in fX$ so that $y = fx$ and $y' = fx'$ for some $x, x' \in X$. By assumption, there is a path $\gamma\colon I \to X$ connecting x and x', and so $f\gamma$ is a path in Y connecting y and y'. □

Corollary 2.1.1 *Connected* and *path connected* are topological properties.

Since quotient maps are continuous surjections, we know quotients preserve (path) connectedness.

Corollary 2.1.2 The quotient of a (path) connected space is (path) connected.

With the right hypothesis, we can go the other way.

Theorem 2.2 Let X be a space and $f : X \to Y$ be a surjective map. If Y is connected in the quotient topology and if each fiber $f^{-1}y$ is connected, then X is connected.

Proof. Let $g : X \to \{0, 1\}$. Since the fibers of f are connected, g must be constant on each fiber of f. Therefore g factors through $f : X \to Y$, and there is a map $\overline{g} : Y \to \{0, 1\}$ that fits into this diagram.

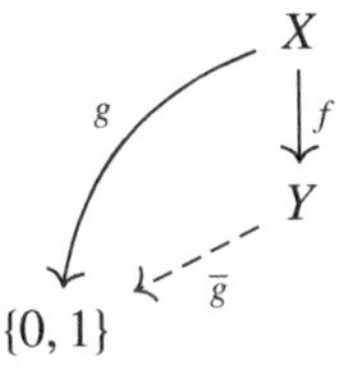

But Y is connected and so $\overline{g}$ must be constant. Therefore $g = \overline{g}f$ is constant. □

Theorem 2.3 Suppose $X = \bigcup_{\alpha \in A} X_\alpha$ and that for each $\alpha \in A$ the space X_α is (path) connected. If there is a point $x \in \bigcap_{\alpha \in A} X_\alpha$ then X is (path) connected.

Proof. Exercise. □

Theorems 2.3 and 2.2 illustrate common strategies in mathematics. Theorem 2.3 involves a space decomposed into a collection of open sets. Information about each open set (they're connected) and information about the intersection (it's nonempty) provides information about the whole space (it's connected). On the other hand, theorem 2.2 involves a space X decomposed into fibers over a base space. Here, information about the base space (it's connected) and information about the fibers (they're connected) provides information about the total space (it's connected). This approach to extending knowledge of parts to knowledge of the whole appears over and over again in mathematics. Something else that commonly appears in mathematics is giving counterexamples to help illuminate definitions.

Example 2.1 The rational numbers $\mathbb{Q}$ are not connected as the continuous map $k : \mathbb{Q} \to \{0, 1\}$ defined by $kx = 0$ if $x < \sqrt{2}$ and $kx = 1$ if $x > \sqrt{2}$ shows. In fact, the rationals are *totally disconnected*, meaning that the only connected subsets are singletons.

This prompts the question: Does $\mathbb{R}$ have *any* connected subsets? If we have a good definition of connected, then an interval ought to be connected. It turns out that there are no other connected subsets of $\mathbb{R}$.

Theorem 2.4 The connected subspaces of $\mathbb{R}$ are intervals.

Proof. Suppose A is a connected subspace of $\mathbb{R}$ that is not an interval. Then there exist $x, y \in A$ such that $x < z < y$ for some $z \notin A$. Thus

$$A = (A \cap (-\infty, z)) \cup (A \cap (z, \infty))$$

is a separation of A into two disjoint nonempty open sets.

Conversely, suppose I is an interval with $I = U \cup V$ where U and V are nonempty, open and disjoint. Then there exist $x \in U$ and $y \in V$, and we may assume $x < y$. Since the set $U' = [x, y) \cap U$ is nonempty and bounded above, $s := \sup U'$ exists by the completeness of $\mathbb{R}$. Moreover, since $x < s \leq y$ and I is an interval, either $s \in U$ or $s \in V$ and so $(s - \delta, s + \delta) \subseteq U$ or $(s - \delta, s + \delta) \subseteq V$ for some $\delta > 0$. If the former holds, then s fails to be an upper bound on U'. If the latter, then $s - \delta$ is an upper bound for U' which is smaller than s. Both lead to a contradiction. □

The completeness of $\mathbb{R}$ was essential in proving the above, which accounts for our use of part (i) of definition 2.1 in the proof in lieu of part (ii). Now that we've proved that the interval $I = [0, 1]$ is connected, we can prove a couple of nice general results about connectedness and path connectedness. Note that since I is connected, the image of any path is connected. That is, if $k: X \to \{0, 1\}$ is a continuous map from a space X, then k is constant along any path $\gamma: I \to X$. Here are a couple of immediate consequences.

Theorem 2.5 Path connected implies connected.

Proof. Suppose X is path connected, and let $k: X \to \{0, 1\}$ be a function. Choose any two points in X. There exists a path connecting them. Since k must be constant on that path, it takes the same value at these two points. Therefore k is constant. □

Theorem 2.6 *Connected* and *path connected* are homotopy invariants.

Proof. Suppose $f: X \to Y$ is a homotopy equivalence, and let $g: Y \to X$ and $h: Y \times I \to Y$ be a homotopy from fg to id_Y.

Suppose that X is connected. To show that Y is connected, let $k: Y \to \{0, 1\}$ be any map, and let $y, y' \in Y$. The map $kf: X \to \{0, 1\}$ must be constant since X is connected, and so $kfgy = kfgy'$. Observe that $h(y, -): I \to Y$ is a path from $h(y, 0) = fgy$ to $h(y, 1) = y$, and so $kfgy = ky$. Also, $h(y', -): I \to Y$ is a path from $h(y', 0) = fgy'$ to $h(y', 1) = y'$, and so $kfgy' = ky'$. Therefore, $ky = ky'$, which implies that k is constant.

Now suppose that X is path connected. Since fX is path connected, we need only worry about its complement. But if $y \in Y \smallsetminus fX$, then $h_y: I \to Y$ is a path from fgy to y. In other words, any point in $Y \smallsetminus fX$ can be connected by a path h_y to a point in fX. Therefore Y is path connected. □

The connectedness of I has other nice consequences. We begin with a fun result we found in Nandakumar and Rao (2012) and Ziegler (2015).

Theorem 2.7 Every convex polygon can be partitioned into two convex polygons, each having the same area and same perimeter.

Proof. Let P be a convex polygon, and first observe that finding a line that bisects the area of P is not difficult. Simply take a vertical line and consider the difference of the area on the left and the right. As the line moves from left to right the difference goes from negative to positive continuously and therefore must be zero at some point. Of course, there was nothing special about a vertical line. There's a line in every direction which bisects P. So start with the vertical line, and consider the difference between the perimeter on the left and the perimeter on the right. Rotate this line in such a way that it always bisects the area of P, and note that the difference between the perimeters switches sign as the line goes halfway around. Therefore there exists a line that cuts P into two convex polygons, both with equal areas and equal perimeters. □

The next result is a special case of *Brouwer's fixed-point theorem*, a landmark theorem in topology.

Theorem 2.8 Every continuous function $f\colon [-1,1] \to [-1,1]$ has a fixed point.

Proof. Suppose $f\colon [-1,1] \to [-1,1]$ is a continuous function for which $fx \neq x$ for all $x \in [-1,1]$. In particular we have $f(-1) > -1$ and $f1 < 1$. Now define a map $g\colon [-1,1] \to \{-1,1\}$ by

$$gx = \frac{x - fx}{|x - fx|}$$

Then g is continuous and $g(-1) = -1$ and $g1 = 1$. But this is impossible since $[-1,1]$ is connected. □

We've just proved the $n = 1$ version of Brouwer's fixed-point theorem which states more generally that for all $n \geq 1$, any continuous function $D^n \to D^n$ must have a fixed point, where n denotes the n-dimensional disk. The result when $n = 2$ is proved in section 6.6.3, where we use a functor called the fundamental group. In fact, we can reprove the $n = 1$ case using a different but closely related functor, π_0.

2.1.2 The Functor π_0

As we hinted earlier in the chapter, there is an assignment $X \mapsto \pi_0 X$ that associates to a space X its set of path components $\pi_0 X$. Now suppose $f\colon X \to Y$ is continuous and $A \subseteq X$ is a path component of X. Then fA is connected and thus contained in a unique path component of Y. Therefore the function $\pi_0 f$ that sends A to the path component containing fA defines a function from $\pi_0 X \to \pi_0 Y$. These data assemble into a functor

$$\mathsf{Top} \xrightarrow{\pi_0} \mathsf{Set}$$

$$\begin{array}{ccc} X & & \pi_0 X \\ {\scriptstyle f}\downarrow & \mapsto & \downarrow{\scriptstyle \pi_0 f} \\ Y & & \pi_0 Y \end{array}$$

Now the fact that functors respect composition when applied to morphisms—often referred to as *functoriality*—makes them quite powerful. To illustrate, let's recast the proof of theorem 2.8 by using the functoriality of π_0. So suppose $f\colon [-1,1] \to [-1,1]$ is continuous. If $fx \neq x$ for any x, then the map $g\colon [-1,1] \to \{-1,1\}$ defined by

$$gx = \frac{x - fx}{|x - fx|} = \begin{cases} -1 & \text{if } x < fx \\ 1 & \text{if } x > fx \end{cases}$$

is continuous, assuming $\{-1,1\}$ is given the discrete topology. So we have a homeomorphism $\{-1,1\} \to \{-1,1\}$ that factors through $[-1,1]$, which is to say it can be written as a composition of the inclusion $i\colon \{-1,1\} \hookrightarrow [-1,1]$ with g.

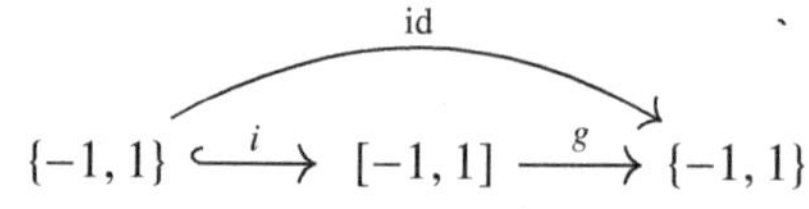

$$\{-1,1\} \overset{i}{\hookrightarrow} [-1,1] \xrightarrow{g} \{-1,1\} \qquad (\text{composite: } \mathrm{id})$$

Applying π_0, we get a diagram of sets

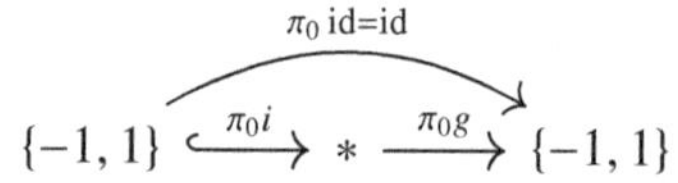

$$\{-1,1\} \xrightarrow{\pi_0 i} * \xrightarrow{\pi_0 g} \{-1,1\} \qquad (\text{composite: } \pi_0\,\mathrm{id} = \mathrm{id})$$

But this is impossible! No map $\{-1,1\} \to *$ can be left invertible, nor can a map $* \to \{-1,1\}$ be right invertible.

2.1.3 Constructions and Connectedness

In chapter 1, we worked through the constructions of new topological spaces from old ones. So far in this chapter, our discussion has centered on two topological properties: connectedness and path connectedness. We've already seen some interactions between these properties and the constructions, but let's systematically run through the the constructions and check whether they preserve connectedness. Quotients do, as stated in corollary 2.1.2. Subspaces don't preserve connectedness: it doesn't take much imagination to come up with an example. Neither do coproducts—the disjoint union of two connected spaces won't be connected—but remember theorem 2.3 if the union is not disjoint. Products, as the next theorem shows, do preserve connectedness.

Theorem 2.9 Let $\{X_\alpha\}_{\alpha\in A}$ be a collection of (path) connected topological spaces. Then $X := \prod_{\alpha\in A} X_\alpha$ is (path) connected.

Proof. We'll prove the theorem for path connected spaces and will leave the rest as an exercise. Suppose X_α is path connected for every $\alpha \in A$, and let $a, b \in X$. Since each X_α is path connected, there exists a path $p_\alpha\colon [0,1] \to X_\alpha$ connecting a_α to b_α. By the universal property of the product topology, the unique function $p\colon [0,1] \to X$ defined by declaring that $\pi_\alpha p = p_\alpha$ for all α is continuous, and moreover p is a path from a to b. □

So we've addressed the direct question for subspaces, quotients, coproducts, and products. Before we move on to other topological properties, there are a couple of other interesting things we can say about connectedness and coproducts.

Every topological space X is partitioned by its connected components $\{X_\alpha\}$. When viewed as a set, X is always equal to the disjoint union of its connected components

$$X = \coprod_\alpha X_\alpha$$

But if we view X as a topological space, it may or may not be homeomorphic to the coproduct of its connected components. For example, the connected components of the rationals $\mathbb{Q}$ are singletons $\{r\}$. But as a topological space, $\mathbb{Q}$ is *not* homeomorphic to $\coprod_{r \in \mathbb{Q}} \{r\}$ (why not?), which is a countable discrete space. A more positive result is the following, whose proof we leave as an exercise.

Theorem 2.10 The following are equivalent.

(i) A space X is the coproduct of its connected components

(ii) The connected components of X are open.

(iii) The quotient space $X/\sim$ of X by its connected components is discrete.

Recall from definition 2.1 that a space is connected if and only if the only maps from it to a two-point discrete space are constant. Let's make this a little more categorical. Observe that for any space X there is exactly one function $X \to *$. And let's think of a two-point discrete space as the coproduct $* \coprod *$. Now if X is connected, then there are precisely two maps $X \to * \coprod *$; namely, the two constant functions: X maps to the first point and X maps to the second point. So the set $\mathsf{Top}(X, * \coprod *)$ is the two-point set, which is canonically isomorphic to $\mathsf{Top}(X, *) \coprod \mathsf{Top}(X, *)$.

However, if X is not connected, then there are *more* than two maps $\mathsf{Top}(X, * \coprod *)$. For example, if $X = [0, 1] \cup [2, 3]$, then there are four functions $X \to * \coprod *$. So the set $\mathsf{Top}(X, * \coprod *)$ is not equal to $\mathsf{Top}(X, *) \coprod \mathsf{Top}(X, *)$. These observations motivate a definition of connectedness that makes sense in *any* category that has coproducts, including Top.

Theorem 2.11 A space X is connected if and only if the functor $\mathsf{Top}(X, -)$ preserves coproducts.

For more information, the categorically minded reader is encouraged to consult the entry on connectedness at the nLab (Stacey et al., 2019).

Wrapping up this brief excursion on constructions, we've seen that connectedness and path-connectedness are preserved by products and quotients but are not preserved by subspaces or coproducts. With that in mind, let's now turn our attention to a *local* version of connectedness.

2.1.4 Local (Path) Connectedness

Definition 2.3 A topological space is *locally connected* (or *locally path connected*) if and only if for each $x \in X$ and every neighborhood $U \subseteq X$ of x, there is a connected (or path connected) neighborhood V of x with $V \subseteq U$.

Example 2.2 Consider the graph of $fx = \sin(1/x)$ where $x > 0$ along with part of the y axis ranging from $(0, -1)$ to $(0, 1)$. This space, called the *topologist's sine curve*, is connected but not path connected. See figure 2.1.

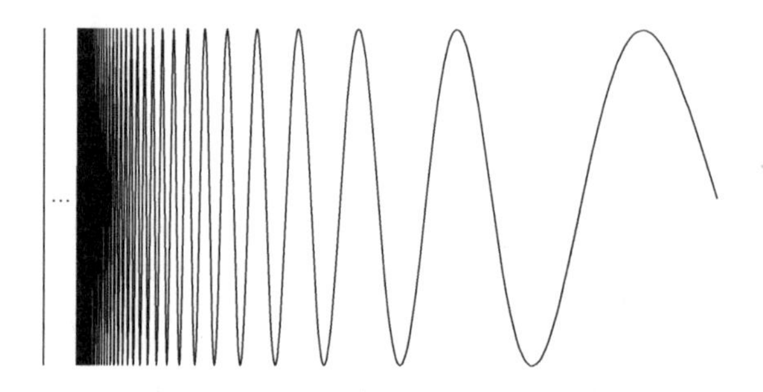

Figure 2.1 The topologist's sine curve

If a space X is locally connected, then the connected components are open, as can be easily verified. This has several consequences. For one, theorem 2.10 implies that locally connected spaces are the coproducts of their connected components. We also have the following.

Theorem 2.12 In any locally path connected topological space, the connected components and path components are the same.

Proof. Exercise. □

Example 2.3 The topologist's sine curve from example 2.2, then, is connected but not locally connected. However, the space $[0, 1] \cup [2, 3]$ is locally connected but not connected.

The previous example illustrates that neither connectedness nor local connectedness implies the other, and the same is true if we replace "connected" with "path connected."

Example 2.4 Let $C = \{\frac{1}{n} \mid n \in \mathbb{N}\} \cup \{0\}$, and set $X = (C \times [0, 1]) \cup ([0, 1] \times \{0\})$. Then X, called the *comb space*, is path connected but not locally path connected. See figure 2.2.

On the other hand, the set $[0, 1] \cup [2, 3]$ in $\mathbb{R}$ with the subspace topology is locally path connected but not path connected.

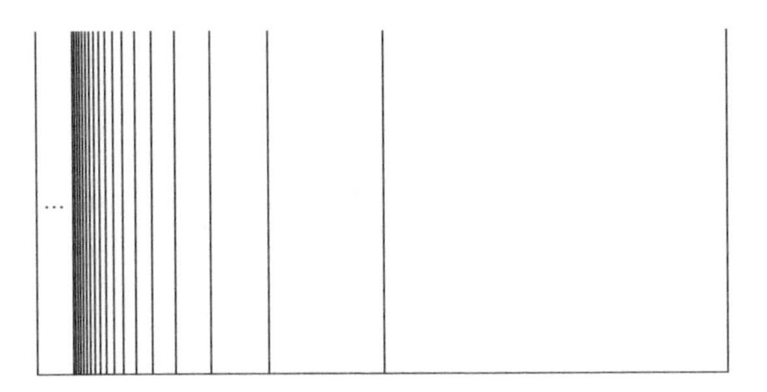

Figure 2.2 The comb space

2.2 Hausdorff Spaces

In the previous section, we discussed connectedness, which in a sense describes when a space can or cannot be separated into nonoverlapping "chunks." The next topological property arises when one seeks for separation at the level of individual points.

Definition 2.4 A space X is *Hausdorff* if and only if for every two points x and y, there exist disjoint open sets U and V with $x \in U$ and $y \in V$.

First, it's good to check that Hausdorff defines a topological property but not a homotopy invariant property. Then we might wonder which constructions preserve the Hausdorff property. One finds that subspaces of Hausdorff spaces are Hausdorff, products of Hausdorff spaces are Hausdorff, and coproducts of Hausdorff spaces are Hausdorff, but quotients of Hausdorff spaces are not necessarily Hausdorff. In fact, quotients of Hausdorff spaces are *the* source of non-Hausdorff spaces throughout the mathematical world. But quotients and Hausdorff spaces do interact well in the following sense.

Theorem 2.13 Every space X is the quotient of a Hausdorff space H.

Proof. Omitted. See Shimrat (1956). □

Example 2.5 Metric spaces are Hausdorff. To see this, let x and y be points in a metric space. If $x \neq y$, then $d := d(x, y) > 0$ and $B(x, \frac{d}{2})$ and $B(y, \frac{d}{2})$ are disjoint open sets separating x and y.

Theorem 2.14 A space X is Hausdorff if and only if the diagonal map $\Delta: X \to X \times X$ is closed.

Proof. Exercise. □

The Hausdorff property interacts with other topological properties in some far-reaching ways. In particular, it gives rise to rich results when combined with compactness.

2.3 Compactness

In this section we introduce the notion of compactness, along with examples and theorems. Admittedly, the proofs in this section have a classical rather than a categorical flavor. But don't fret. Instead, we encourage you to eagerly anticipate chapter 5 where we'll revisit compact Hausdorff spaces in great categorical detail.

2.3.1 Definitions, Theorems, and Examples

Definition 2.5 A collection $\mathcal{U}$ of open subsets of a space X is called an *open cover* for X if and only if the union of sets in $\mathcal{U}$ contains X. The space X is *compact* if and only if every open cover of X has a finite subcover.

Theorem 2.15 If X is compact and $f: X \to Y$ is continuous, then fX is compact.

Proof. Exercise. □

Corollary 2.15.1 *Compactness* is a topological property.

One way to think of compact spaces is that they are somehow small—not in terms of cardinality but in terms of roominess. For example, if you squeeze an infinite set of points into the unit interval, they'll get cramped—for any $\varepsilon > 0$, there are two points that are less than ε apart. But it's easy to fit an infinite number of points in the real line so that they're all spread out. Indeed, the unit interval is compact while the real line is not. This idea is summarized in the next theorem. First, a piece of terminology. A point x is called a *limit point* of a space X if every neighborhood of x contains a point of $X \setminus \{x\}$.

The Bolzano-Weierstrass Theorem Every infinite set in a compact space has a limit point.

Proof. Suppose that F is an infinite subset with no limit points. If x is not a limit point of F and $x \notin F$, there is an open set U_x around x that misses F. If x is not a limit point of F and $x \in F$, then there is an open set U_x with $U_x \cap F = \{x\}$. Then $\{U_x\}_{x \in X}$

is an open cover of X. Notice that there can be no finite subcover $U_{x_1}, \ldots, U_{x_n}$ since $(U_{x_1} \cup \cdots \cup U_{x_n}) \cap F = \{x_1, \ldots, x_n\}$, and cannot contain the infinite set F. □

Example 2.6 Note that compactness is not necessary in the previous theorem, as there exist noncompact spaces for which every infinite subset has a limit point. For instance, take $\mathbb{R}$ with topology $\{(x, \infty) : x \in \mathbb{R}\}$ together with $\varnothing$ and $\mathbb{R}$. This space is not compact, but any set (infinite or not) has a limit point (infinitely many, in fact).

In general, directly checking if a space is compact can be tricky. The following definition sets the stage for an alternate criterion, as described in the next theorem.

Definition 2.6 Let $\mathcal{S}$ be a collection of sets. We say that $\mathcal{S}$ has the *finite intersection property* if and only if for every finite subcollection $A_1, \ldots, A_n \subseteq \mathcal{S}$, the intersection $A_1 \cap \cdots \cap A_n \neq \varnothing$. We abbreviate the finite intersection property by FIP.

Theorem 2.16 A space X is compact if and only if every collection of closed subsets of X with the FIP has nonempty intersection.

Proof. Exercise. □

Here's yet another way to check for compactness.

Theorem 2.17 Closed subsets of compact spaces are compact.

Proof. Let X be compact with $C \subseteq X$ closed and suppose $\mathcal{U} = \{U_\alpha\}_{\alpha \in A}$ is an open cover of C. Then $X \smallsetminus C$ together with $\mathcal{U}$ forms an open cover of X. Since X is compact, there are finitely many sets $\{U_i\}_{i=1}^n$ in $\mathcal{U}$, possibly together with $X \smallsetminus C$, which covers X. Thus $\{U_i\}_{i=1}^n$ is a finite subcover for C. □

Now we're ready to see how compactness and the Hausdorff property interact. To start, compact subsets of Hausdorff spaces are quite nice—they can be separated from points by open sets.

Theorem 2.18 Let X be Hausdorff. For any point $x \in X$ and any compact set $K \subseteq X \smallsetminus \{x\}$ there exist disjoint open sets U and V with $x \in U$ and $K \subseteq V$.

Proof. Let $x \in X$ and let $K \subsetneq X$ be compact. For each $y \in K$, there are disjoint open sets U_y and V_y with $x \in U_y$ and $y \in V_y$. The collection $\{V_y\}$ is an open cover of K; hence there is a finite subcover $\{V_1, \ldots, V_n\}$. Let $U = U_1 \cap \cdots \cap U_n$ and $V = V_1 \cup \cdots \cup V_n$. Then U and V are disjoint open sets with $x \in U$ and $K \subseteq V$. □

This theorem quickly gives rise to two important corollaries.

Corollary 2.18.1 Compact subsets of Hausdorff spaces are closed.

Proof. Exercise. □

Corollary 2.18.2 If X is compact and Y is Hausdorff, then every map $f: X \to Y$ is closed.[1] In particular,

- if f is injective, then it is an embedding;
- if f is surjective, then it is a quotient map;
- if f is bijective, then it is a homeomorphism.

Proof. Let $f: X \to Y$ be a map from a compact space to a Hausdorff space, and let $C \subseteq X$ be closed. Then C is compact, so fC is compact, so fC is closed. □

As you'll recall from example 1.13, not every continuous bijection $f: X \to Y$ is a homeomorphism. The previous corollary guarantees us that such maps are homeomorphisms whenever X is compact and Y is Hausdorff.

2.3.2 Constructions and Compactness

As with our discussion on connectedness, we are also interested in the preservation of compactness under the four constructions: subspaces, quotients, products, and coproducts. Subspaces of compact spaces are not compact in general, but we saw in theorem 2.17 that closed subspaces of compact spaces are compact. You'll also realize that we've proved that quotients of compact spaces are compact. Coproducts of compact spaces are certainly not compact—just look at the coproduct of infinitely many copies of a point. But what about products? There are a few interesting things to explain here, and we'll start with *Tychonoff's theorem* and some of its corollaries.

Tychonoff's Theorem 1 The product of compact spaces is compact.

Proof. See section 3.4. □

Corollary 2.18.3 (Heine-Borel Theorem) A subset of $\mathbb{R}^n$ is compact if and only if it is closed and bounded.

Proof. Suppose that $K \subset \mathbb{R}^n$ is compact. Since the cover of K consisting of open balls centered at the origin of all possible radii must have a finite subcover, K must be bounded. Since $\mathbb{R}^n$ is Hausdorff and all compact subsets of a Hausdorff space must be closed, K is closed.

Conversely (and this is the part that uses the Tychonoff theorem), suppose that $K \subset \mathbb{R}^n$ is closed and bounded. Since K is bounded, the projection of K onto the ith coordinate is bounded; that is, for each i there's an interval $[a_i, b_i]$ containing $\pi_i K$. Then $K \subseteq [a_1, b_1] \times [a_2, b_2] \times \cdots \times [a_n, b_n]$. Since each set $[a_i, b_i]$ is compact, the Tychonoff theorem implies

[1] The map f is *closed* if fC is closed whenever $C \subseteq X$ is closed. See exercise 1.14 at the end of chapter 1.

that the product $[a_1, b_1] \times [a_2, b_2] \times \cdots \times [a_n, b_n]$ is compact. Since any closed subset of a compact space is compact, we conclude that K is compact. □

Corollary 2.18.4 Continuous functions from compact spaces to $\mathbb{R}$ have both a global maximum and a global minimum.

Proof. Exercise. □

The characterization of compact subsets of $\mathbb{R}^n$ as closed and bounded may be familiar from analysis, but recall that *bounded* is *not* a topological property! For example, there is a homeomorphism of topological spaces $\mathbb{R} \cong (0, 1)$, yet $\mathbb{R}$ is not a bounded metric space while $(0, 1)$ is. It's also not a homotopy invariant, and neither is compactness.

Example 2.7 Like any space whose underlying set is finite, the one-point set $*$ is compact. Since $\mathbb{R}$ is not compact but is homotopy equivalent to $*$, we see that compactness is not a homotopy invariant.

Finally, we have the so-called *Tube Lemma*, which isn't a corollary of Tychonoff, but it does concern compact sets and products. First, here's an example.

Example 2.8 Let U be the interior of the triangle with corners $(0, 0)$, $(1, 0)$, and $(1, 1)$—

$$U := \{(x, y) \subset \mathbb{R}^2 \mid 0 < x < 1,\ 0 < y < x\}$$

—and consider the set $A \times \left\{\frac{1}{2}\right\}$ where A is the interval $A = \left(\frac{1}{2}, 1\right)$. Then $A \times \left(\frac{1}{2} - \varepsilon, \frac{1}{2} + \varepsilon\right)$ is not contained in U for any $\varepsilon > 0$. But if A were compact....

The Tube Lemma For any open set $U \subseteq X \times Y$ and any set $K \times \{y\} \subseteq U$ with $K \subseteq X$ compact, there exist open sets $V \subseteq X$ and $W \subseteq Y$ with $K \times \{y\} \subseteq V \times W \subseteq U$.

Proof. For each point $(x, y) \in K \times \{y\}$, there are open sets $V_x \subseteq X$ and $W_x \subseteq Y$ with $(x, y) \in V_x \times W_x \subseteq U$. Then, $\{V_x\}_{x \in K}$ is an open cover of K; take a finite subcover $\{V_1, \ldots, V_n\}$. Then $V = V_1 \cup \cdots \cup V_n$ and $W = W_1 \cap \cdots \cap W_n$ are open sets with $K \times \{y\} \subseteq V \times W \subseteq U$. □

We now close by briefly mentioning the local version of compactness.

2.3.3 Local Compactness

We will define local compactness by way of saying that "spaces that are *locally compact* are spaces whose neighborhoods look like neighborhoods of compact spaces."

Definition 2.7 A space X is *locally compact* if and only if for every point $x \in X$ there exists a compact set K and a neighborhood U with $x \in U \subseteq K$.

Example 2.9 Every compact space is locally compact, as is every discrete space. Also, $\mathbb{R}^n$ is locally compact; however, the real line with the lower limit topology $\mathcal{T}_{\text{ll}}$ (example 1.3) is not, as the reader can verify.

Note that the image of a locally compact space need not be locally compact. For example, consider the map $\mathrm{id}\colon (\mathbb{R}, \mathcal{T}_{\text{discrete}}) \to (\mathbb{R}, \mathcal{T}_{\text{ll}})$. Nonetheless, locally compact is a topological property, as one can verify. For Hausdorff spaces, local compactness is much stronger.

Theorem 2.19 Suppose X is locally compact and Hausdorff. Then for every point $x \in X$ and every neighborhood U of x, there exists a neighborhood V of x such that the closure $\overline{V}$ is compact and $x \in V \subseteq \overline{V} \subseteq U$.

Proof. This is a corollary of theorem 2.18 and the definition of local compactness. □

Lastly, we mention that the product and quotient topologies are not compatible in the sense of exercise 1.11 at the end of chapter 1, but the hypothesis of locally compact and Hausdorff makes the situation much better.

Theorem 2.20 If $X_1 \twoheadrightarrow Y_1$ and $X_2 \twoheadrightarrow Y_2$ are quotient maps and Y_1 and X_2 are locally compact and Hausdorff, then $X_1 \times X_2 \twoheadrightarrow Y_1 \times Y_2$ is a quotient map.

Proof. We postpone the proof until theorem 5.7 in chapter 5. □

Exercises

1. Prove that the two items in definition 2.1 are indeed equivalent.

2. A map $X \to Y$ is *locally constant* if for each $x \in X$ there is an open set U with $x \in U$ and $f|_U$ constant. Prove or disprove: if X is connected and Y is any space, then every locally constant map $f: X \to Y$ is constant.

3. Show that every countable metric space with at least two points must be disconnected. Construct a topological space with more than two elements that is both countable and connected.

4. In a variation of the topology on $\mathbb{Z}$ in example 1.5, consider the natural numbers $\mathbb{N}$ with topology generated by the basis

$$\{ak + b \mid k \in \mathbb{N} \text{ and } a, b \in \mathbb{N} \text{ are relatively prime}\}$$

Prove that $\mathbb{N}$ with this topology is connected (Golomb, 1959).

5. Let $\{X_\alpha\}_\alpha$ be a collection of spaces. Prove that $\pi_0 \prod X_\alpha \cong \prod \pi_0 X_\alpha$. Note: the special case $\pi_0 X_\alpha = *$ for all α is the statement that the product of path connected spaces is path connected.

6. Provide a proof of theorem 2.10.

7. Prove that a space X is connected if and only if the functor $\mathsf{Top}(X, -)$ preserves coproducts.

8. Show that $\mathbb{Q} \subseteq \mathbb{R}$ with the subspace topology is not locally compact.

9. Prove that the product of two locally compact Hausdorff spaces is locally compact Hausdorff.

10. Define a space X to be *pseudocompact* if and only if every real valued function on X is bounded. Prove that if X is compact, then X is pseudocompact, and give an example of a pseudocompact space that is not compact.

11. Give examples showing that locally compact is not preserved by subspaces, quotients, or products.

12. Let $\mathcal{U}$ be an open cover of a compact metric space X. Show that there exists an $\varepsilon > 0$ such that for every $x \in X$, the set $B(x, \varepsilon)$ is contained in some $U \in \mathcal{U}$. Such an ε is called a *Lebesgue number for* $\mathcal{U}$.

13. Show that $\mathbb{Z}$ endowed with the arithmetic progression topology of example 1.5 is not locally compact.

14. Suppose (X, d) is a compact metric space and $f: X \to X$ is an isometry; that is, for all $x, y \in X$, $d(x, y) = d(fx, fy)$. Prove f is a homeomorphism.

15. Let X be a space and suppose $A, B \subseteq X$ are compact. Prove or disprove:

a) $A \cap B$ is compact.

b) $A \cup B$ is compact.

If a statement is false, find a sufficient condition on X which will cause it to be true.

16. Let $B = \left\{\{x_n\} \in l^2 \mid \sum_{n=1}^{\infty} x_n^2 \leq 1\right\}$ be the closed unit ball in l^2, where l^2 is the space defined in example 1.8 of chapter 1. Show that B is not compact.

17. Prove that if Y is compact, then for any space X the projection $X \times Y \to X$ is a closed map. Give an example of spaces X and Y for which the projection $X \times Y \to X$ is not closed.

18. Show that the product of Hausdorff spaces is Hausdorff. Give an example to show that the quotient of a Hausdorff space need not be Hausdorff.

19. If X is any set and Y is Hausdorff, then a subset $A \subseteq \mathsf{Top}(X, Y)$ has compact closure in the product topology if and only if for each $x \in X$, the set $A_x = \{fx \in Y \mid f \in A\}$ has compact closure in Y.

20. For any map $f \colon X \to Y$, the set $\Gamma = \{(x, y) \in X \times Y \mid y = fx\}$ is called the *graph of f*. Suppose now that X is any space and Y is compact Hausdorff. Prove that Γ is closed if and only if f is continuous. Is the compactness condition necessary? (This is called the *closed graph theorem.*)

21. Let X be a Hausdorff space with $f \colon X \to Y$ a continuous closed surjection such that $f^{-1}y$ is compact for each $y \in Y$. Prove that Y is Hausdorff.

22. Prove or disprove: if $f \colon X \to Y$ is a continuous bijection and X is Hausdorff, then Y must be Hausdorff.

23. Prove or disprove: X is Hausdorff if and only if

$$\{(x, x, \ldots) \in X^{\mathbb{N}} \mid x \in X\}$$

is closed in $X^{\mathbb{N}}$.

24. Topologies that are compact and Hausdorff are nicely balanced. Take for an example $[0, 1]$.

a) Prove that if $\mathcal{T}$ is any topology on $[0, 1]$ finer than the ordinary one, then $[0, 1]$ cannot be compact in the topology $\mathcal{T}$.

b) Prove that if $\mathcal{T}$ is any topology on $[0, 1]$ coarser than the usual one, then $[0, 1]$ cannot be Hausdorff in the topology $\mathcal{T}$.

3 Limits of Sequences and Filters

The Axiom of Choice is obviously true, the well-ordering theorem is obviously false; and who can tell about Zorn's Lemma?
—Jerry Bona (Schechter, 1996)

Introduction. Chapter 2 featured various properties of topological spaces and explored their interactions with a few categorical constructions. In this chapter we'll again discuss some topological properties, this time with an eye toward more fine-grained ideas. As introduced early in a study of analysis, properties of nice topological spaces X can be detected by sequences of points in X. We'll be interested in some of these properties and the extent to which sequences suffice to detect them. But take note of the adjective "nice" here. What if X is any topological space, not just a nice one? Unfortunately, sequences are not well suited for characterizing properties in arbitrary spaces. But all is not lost. A sequence can be replaced with a more general construction—*a filter*—which is much better suited for the task. In this chapter we introduce filters and highlight some of their strengths.

Our goal is to spend a little time inside of spaces to discuss ideas that may be familiar from analysis. For this reason, this chapter contains less category theory than others. On the other hand, we'll see in section 3.3 that filters are a bit like functors and hence like generalizations of points. This perspective thus gives us a coarse-grained approach to investigating fine-grained ideas. We'll go through some of these basic ideas—closure, limit points, sequences, and more—rather quickly in sections 3.1 and 3.2. Later in section 3.2 we'll see exactly why sequences don't suffice to detect certain properties in all spaces. We'll also discover those "nice" spaces for which they do. Section 3.3 introduces filters with some examples and results about them. These results include the claim that filters, unlike sequences, do suffice to characterize certain properties. Finally, in section 3.4 we'll use filters to share a delightfully short proof of Tychonoff's theorem.

3.1 Closure and Interior

Here are a few basic definitions, which may be familiar from analysis. Given any subset B of a space X, two topological constructions suggest themselves. There is the *closure* $\overline{B}$ which is the smallest closed set containing B, and there is the *interior* B° which is the largest open set contained in B. When $\overline{B} = X$, we say B is *dense* in X. If $\left(\overline{B}\right)^\circ = \varnothing$, we say B is *nowhere dense*.

Notice that the definition of a topology guarantees that the interior and closure exist. For example, because a topology is closed under arbitrary unions, the interior B° is precisely the union of all open subsets of B. Contrast this with the ideas of a "largest closed set" contained in B and a "smallest open set" containing B, which might not exist.

A point x is called a *limit point* of a set B if every open set around x contains a point of $B \setminus \{x\}$. The closure $\overline{B}$ consists of B together with all of its limit points. A point x is called a *boundary point* of B if every open set containing x contains both a point in B and a point in the complement of B.

Limit points help to understand closures and interiors. So, taking a cue from analysis, let's turn to a study of *sequences* in an attempt to characterize limit points.

3.2 Sequences

Definition 3.1 Let X be a topological space. A *sequence* in X is a function $x\colon \mathbb{N} \to X$. We usually write x_n for $x(n)$ and may denote the sequence by $\{x_n\}$. A sequence $\{x_n\}$ *converges* to $z \in X$ if and only if for every open set U containing z, there exists an $N \in \mathbb{N}$ so that if $n \geq N$ then $x_n \in U$. When $\{x_n\}$ converges to $z \in X$ we'll write $\{x_n\} \to z$. A *subsequence* of a sequence x is the composition xk where $k\colon \mathbb{N} \to \mathbb{N}$ is an increasing injection. We'll write x_{k_i} for $xk(i)$ and denote the subsequence by $\{x_{k_i}\}$.

Here are a few examples.

Example 3.1 Let $A = \{1, 2, 3\}$ with the topology $\mathcal{T} = \{\varnothing, \{1\}, \{1, 2\}, A\}$. The constant sequence $1, 1, 1, 1, \ldots$ converges to 1. It also converges to 2 and to 3.

Example 3.2 Consider $\mathbb{Z}$ with the cofinite topology. For any $m \in \mathbb{Z}$, the constant sequence $m, m, m, \ldots$ converges to m and only to m. Indeed if $l \neq m$, then the set $\mathbb{R} \setminus m$ is an open set around l containing no elements of the sequence.

However, the sequence $\{n\} = 1, 2, 3, 4, \ldots$ converges to m for every $m \in \mathbb{Z}$. To see this, let m be any integer, and let U be a neighborhood of m. Since $\mathbb{Z} \setminus U$ is finite, there can only be finitely many natural numbers in $\mathbb{Z} \setminus U$. Let N be larger than the greatest natural number in $\mathbb{Z} \setminus U$. Then $n \in U$ whenever $n \geq N$, proving that $\{n\} \to m$.

Example 3.3 Consider $\mathbb{R}$ with the usual topology. If $\{x_n\} \to x$, then $\{x_n\}$ does not converge to any number $y \neq x$. To prove it, note we can always find disjoint open sets U and V with $x \in U$ and $y \in V$. (We can be explicit if necessary: $U = (x - c, x + c)$ and $V = (y - c, y + c)$ where $c = \frac{1}{2}|x - y|$.) Then there is a number N so that $x_n \in U$ for all $n \geq N$. Since $U \cap V = \varnothing$, the set V cannot contain any x_n for $n \geq N$, and hence $\{x_n\}$ does not converge to y.

As we'll see below, sequences can be used to detect certain properties of spaces, subsets of spaces, and functions between spaces. But before continuing, it will be helpful to first introduce a couple more topological properties—two of the so-called "separation" axioms.

Definition 3.2 We say

(i) A topological space X is T_0 if and only if for every pair of points $x, y \in X$ there exists an open set containing one, but not both, of them.

(ii) A topological space X is T_1 if and only if for every pair of points $x, y \in X$ there exist open sets U and V with $x \in U$, $y \in V$ with $x \notin V$ and $y \notin U$.

We could have added a third property to the list. A space X with the property that for every pair of points $x, y \in X$ there exist open sets U and V with $x \in U$, $y \in V$ with $U \cap V = \varnothing$ is sometimes called T_2, but we've already named the property *Hausdorff*, after Felix Hausdorff who originally used the axiom in his definition of "neighborhood spaces" (Hausdorff and Aumann, 1914). Note that T_0, T_1, and T_2 all define topological properties.

Here are a few theorems about sequences that might evoke familiar results from analysis. Some of the proofs are left as exercises. You'll want to keep the examples above in mind.

Theorem 3.1 A space X is T_1 if and only if for any $x \in X$, the constant sequence $x, x, x, \ldots$ converges to x and only to x.

Proof. Suppose X is T_1 and $x \in X$. It's clear that $x, x, x, \ldots \to x$. Let $y \neq x$. Then there exists an open set U with $y \in U$ and $x \notin U$. Therefore, $x, x, x, \ldots$ cannot converge to y.

For the converse, suppose X is not T_1. Then there exist two distinct points x and y for which every open set around y contains x. Thus $x, x, x, \ldots \to y$. □

Theorem 3.2 If X is Hausdorff, then sequences in X have at most one limit.

Proof. Let X be Hausdorff and suppose $\{x_n\}$ is a sequence such that $\{x_n\} \to x$ and $\{x_n\} \to y$. If $x \neq y$, then there are disjoint open sets U and V with $x \in U$ and $y \in V$. Since $\{x_n\} \to x$ there is a number N so that $x_n \in U$ for all $n \geq N$. Since $\{x_n\} \to y$ there is a number K so that $x_n \in U$ for all $n \geq K$. Let $M = \max\{N, K\}$. Since $M \geq N$ and $M \geq K$ we have $x_M \in U$ and $x_M \in V$, contradicting the fact that U and V are disjoint. □

Theorem 3.3 If $\{x_n\}$ is a sequence in A that converges to x, then $x \in \overline{A}$.

Proof. Exercise. □

Theorem 3.4 If $f: X \to Y$ is continuous, then for all sequences $\{x_n\}$ that converge to x in X, the sequence $\{fx_n\}$ converges to fx in Y.

Proof. Exercise. □

You'll notice that theorem 3.1 is an if-and-only-if theorem that characterizes the T_1 property with a statement about sequences. So you might wonder if sequences are also enough to characterize Hausdorff spaces, closed sets, and continuous functions. That is, do theorems 3.2, 3.3, and 3.4 have if-and-only-if-versions, too? The answer is no.

Example 3.4 Sequences don't suffice to detect Hausdorff spaces. Consider $\mathbb{R}$ with the cocountable topology. This space is not Hausdorff, and yet convergent sequences have unique limits.

Example 3.5 Sequences don't suffice to detect closed sets. Let

$$X = [0, 1]^{[0,1]} := \{\text{functions } f: [0, 1] \to [0, 1]\}$$

with the product topology, and let A be the subset of X consisting of functions whose graphs are "sawtooths" with vertices on the x axis at a finite number of points $\{0, r_1, \ldots, r_n, 1\}$ and spikes of height 1, as in figure 3.1. The zero function is in $\overline{A}$, but there is no sequence $\{f_n\}$ in A converging to it.

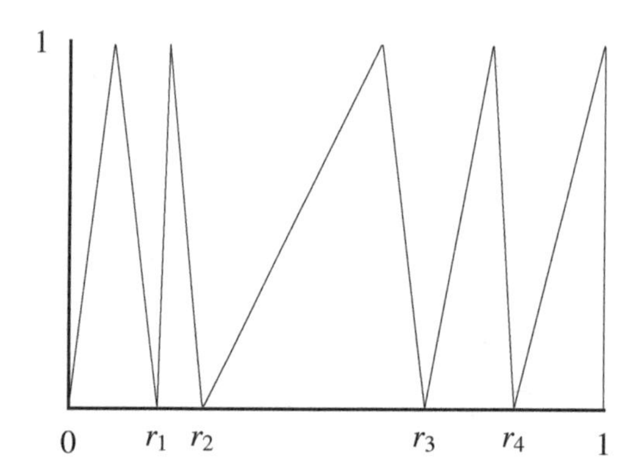

Figure 3.1 A sawtooth function

Example 3.6 Sequences don't suffice to detect continuous functions. Let $X = [0, 1]^{[0,1]} :=$ $\{\text{functions } f: [0, 1] \to [0, 1]\}$ with the product topology, and let Y be the subspace of X consisting of integrable functions. The function $I: Y \to \mathbb{R}$ defined by $If = \int_0^1 f$ is not a continuous function but $\{If_n\} \to If$ whenever $\{f_n\} \to f$.

Sequences don't suffice to detect the Hausdorff property, closure, and continuity in these examples because the spaces in question have too many open sets around each point for their topological properties to be adequately probed by sequences. But for spaces without too many open sets around each point, sequences do suffice to characterize their properties. These spaces are called *first countable.*

Definition 3.3 Let X be a space. A collection of open sets $\mathcal{B}$ is called a *neighborhood base* for $x \in X$ if and only if for every open set O containing x, there exists an open set $U \in \mathcal{B}$ with $x \in U \subseteq O$. A space X is called *first countable* if and only if every point has a countable neighborhood base. A space X is called *second countable* if and only if it has a countable basis.

Example 3.7 By definition, the set $\mathcal{T}_x$ of open neighborhoods of a point x in a space X is a neighborhood base for x.

Example 3.8 Every metric space is first countable since the open balls around x of radius $1, \frac{1}{2}, \frac{1}{3}, \ldots$ form a countable neighborhood base.

Example 3.9 An n-dimensional *manifold* is a second countable Hausdorff topological space with the property that every point has a neighborhood homeomorphic to $\mathbb{R}^n$.

Two examples of nonfirst countable spaces were given in example 3.4, example 3.5, and example 3.6; namely, $\mathbb{R}$ with the cocountable topology and $[0, 1]^{[0,1]}$ with the product topology. But in first countable spaces such as metric spaces, sequences *do* suffice to characterize separation, closure, and continuity properties. In other words, theorem 3.2, theorem 3.3, and theorem 3.4 do have if-and-only-if versions in this special context.

Theorem 3.5 Let X be a first countable space. Then X is Hausdorff if and only if every sequence has at most one limit.

Proof. Suppose that X is first countable. If X is not Hausdorff, there exist points x and y that cannot be separated by open sets. Let $U_1, U_2, \ldots$ be a neighborhood base of x and $V_1, V_2, \ldots$ be a neighborhood base for y. For every n choose a point $x_n \in U_n \cap V_n \neq \varnothing$. The sequence $\{x_n\}$ has a subsequence that converges to x and to y. □

Theorem 3.6 Let X be a first countable space and let $A \subseteq X$. A point $x \in \overline{A}$ if and only if there exists a sequence $\{x_n\}$ in A with $\{x_n\} \to x$.

Proof. Exercise. □

Theorem 3.7 Suppose X and Y are first countable. A function $f \colon X \to Y$ is continuous if and only if for every sequence $\{x_n\}$ in X with $\{x_n\} \to x$, the sequence $\{fx_n\} \to fx$.

Proof. Exercise. □

The reason sequences characterize separation, closure, and continuity in first countable spaces but not in arbitrary spaces is simply because sequences are countable. If, however, we want the previous three if-and-only-if theorems to hold in a wider context, then we'll want a generalization of sequences—one that more accurately reflects the range of possibilities for convergence. This generalization is the topic of the next section. To help introduce the ideas, let's make a simple observation.

Given a sequence $\{x_n\}$ in a space X, consider the collection of sets the sequence is eventually inside of:

$$\mathcal{E}_{x_n} := \{A \subseteq X \mid \text{there exists an } N \text{ so that } x_n \in A \text{ for all } n \geq N\} \tag{3.1}$$

Now here's the key:

> The sequence $\{x_n\}$ converges to a point x if and only if the neighborhood base $\mathcal{T}_x$ is contained in $\mathcal{E}_{x_n}$.

Understanding convergence, therefore, amounts to understanding the set $\mathcal{E}_{x_n}$. This immediately suggests how one might attempt to generalize sequences: abstract the notion of "the set of sets the sequence is eventually inside of." Cartan (1937b) did just that in 1937, when he introduced an appropriate generalization of sequences well suited to studying convergence: *filters*. As we'll soon see, filters are just what's needed to obtain the three analogous if-and-only-if theorems for *all* spaces, not just first countable ones.

3.3 Filters and Convergence

A filter is like an algebraic road map with (perhaps rough) directions to points or places in a space. More precisely, it is a certain subset of a poset (a partially ordered set). In this chapter, the poset we focus on is the powerset of a set X. That is, we consider *filters of subsets of a given set*. Here is the definition.

Definition 3.4 A *filter* on a set X is a collection $\mathcal{F} \subseteq 2^X$ that is

(i) downward directed: $A, B \in \mathcal{F}$ implies there exists $C \in \mathcal{F}$ such that $C \subseteq A \cap B$
(ii) nonempty: $\mathcal{F} \neq \varnothing$
(iii) upward closed: $A \in \mathcal{F}$ and $A \subseteq B$ implies $B \in \mathcal{F}$

An additional property is often useful:

(iv) proper: there exists $A \subseteq X$ such that $A \notin \mathcal{F}$

So a filter on X is a downward directed, nonempty, upward closed subset of the powerset 2^X. A couple of comments are in order. Being downward directed and upward closed implies that filters are closed under finite intersections. We can also rephrase properness as the requirement that $\varnothing \notin \mathcal{F}$. For example, 2^X is itself a filter and is well named the *improper filter* on X. A set that is only downward directed and nonempty is called a *filterbase*. Any filterbase generates a filter; just take the upward closure of the base.

In chapter 0, it was noted that any poset can be viewed as a category; objects are elements in the set, and morphisms are provided by the partial order. So there's hope that filters have a categorical description. Indeed they do. It starts with the observation that the poset 2^X has the property that every pair of elements $A, B \in 2^X$ has a *meet* (a greatest lower bound), namely their intersection $A \cap B$. We can define another poset also having this property. Consider the two-element poset $2 := \{0 \leq 1\}$. For $a, b \in 2$ define their meet $a \wedge b$ by

$$0 \wedge 0 = 0 \qquad 0 \wedge 1 = 0 \qquad 1 \wedge 0 = 0 \qquad 1 \wedge 1 = 1$$

Every monotone function $f\colon 2^X \to 2$ that respects this structure—that is, that satisfies $f(A \cap B) = fA \wedge fB$—defines a filter, namely the preimage $f^{-1}1$. Verifying this claim is a simple exercise. In the language of order theory, f is called a *meet-semilattice homomorphism*. In the language of category theory, it is called a *continuous functor*.

Indeed, the posets 2^X and 2 are categories, and a functor between them is precisely a monotone function. We'll see in chapter 4 that a meet is an example of a more general categorical construction called a *limit*, and a functor that respects limits is, with inspiration from theorem 3.4, called *continuous*. Filters thus arise from continuous functors $2^X \to 2$.

Example 3.10 For any set X, the *trivial filter* $\mathcal{F} = \{X\}$ is a proper filter. More generally, for any nonempty subset $A \subseteq X$, the set of all sets containing A is a proper filter.

Example 3.11 Another example of a proper filter is the *eventuality filter* $\mathcal{E}_{x_n}$ associated to the sequence $\{x_n\}$ from equation 3.1. It's aptly named since it's the collection of all sets that the sequence eventually remains in.

Example 3.12 The cofinite subsets of a set X

$$\mathcal{F} := \{A \subseteq X \mid X \setminus A \text{ finite}\}$$

define a filter called the *Fréchet filter*. If X is infinite, then the Frećhet filter is proper.

Example 3.13 Given a topological space $(X, \mathcal{T})$, the open neighborhoods $\mathcal{T}_x$ of a point x form a filterbase, though they generally do not form a filter. The reason is simply that (usually) not every set containing an open neighborhood of x is open. But there is some ambiguity in the mathematics community on this point. Kelley (1955) defines "a neighborhood of x" to mean exactly "a set containing an open set containing x." Others, such as Munkres (2000), prefer all neighborhoods be open neighborhoods. In our language, the filterbase $\mathcal{T}_x$ generates the filter of (not necessarily open) neighborhoods of x.

We began our discussion about filters with an observation about convergence: a sequence in a topological space converges if and only if its eventuality filter contains the filterbase $\mathcal{T}_x$. This motivates the following definition.

Definition 3.5 A filter $\mathcal{F}$ on a topological space $(X, \mathcal{T})$ *converges* to x if and only if $\mathcal{F}$ refines $\mathcal{T}_x$, that is if $\mathcal{T}_x \subseteq \mathcal{F}$. When $\mathcal{F}$ converges to x we'll write $\mathcal{F} \to x$.

Example 3.14 The real-valued sequence $\{x_n\} := \{1, -1, \frac{1}{2}, -1, \frac{1}{4}, -1, \frac{1}{8}, \ldots\}$ does not converge, whereas the subsequence $\{x_{2n}\} = \{1, \frac{1}{2}, \frac{1}{4}, \frac{1}{8}, \ldots\}$ does. We can see this by reasoning with eventuality filters, which isn't so different from reasoning with sequences. Here's the thing to notice: the eventuality filter $\mathcal{E}_{x_{2n}}$ is the set of all subsets $A \subset \mathbb{R}$ for which there exists an N so that $\frac{1}{2^n} \in A$ for all $n \geq N$. It's straightforward to check that $\mathcal{E}_{x_{2n}} \to 0$. However, the eventuality filter $\mathcal{E}_{x_n}$ has the same description as $\mathcal{E}_{x_{2n}}$, except each A must also include -1. So $\mathcal{E}_{x_n} \subseteq \mathcal{E}_{x_{2n}}$.

This example illustrates that passing to a subsequence increases the size of an eventuality filter since the membership condition is weaker. At the extreme end of this, the improper filter $\mathcal{F} = 2^X$ converges to *every point* in X! (And yet it is not the eventuality filter of any sequence.)

At this juncture, you'll recall our earlier claim that filters suffice to give a characterization of the Hausdorff property, closure, and continuity. The proofs of the first two are now available.

Theorem 3.8 A space is Hausdorff if and only if limits of convergent proper filters are unique.

Proof. Suppose X is Hausdorff and that a proper filter $\mathcal{F}$ converges to both x and y with $x \neq y$. Then there are open neighborhoods U of x and V of y with $U \cap V = \varnothing$. By convergence, $U, V \in \mathcal{F}$. Since $\mathcal{F}$ is a filter, $\varnothing = U \cap V \in \mathcal{F}$, which contradicts $\mathcal{F}$ being proper.

However, if X is not Hausdorff then there are two distinct points x and y that cannot be separated by open sets. Let

$$\mathcal{B} = \{U \cap V \mid x \in U, y \in V \text{ with } U \text{ and } V \text{ open}\}$$

Note that $\mathcal{B}$ is downward directed and nonempty, so it is a filterbase. The filter $\mathcal{F}$ generated by the collection $\mathcal{B}$ converges to both x and y. □

Theorem 3.9 Let X be a space with $A \subseteq X$. A point $x \in \overline{A}$ if and only if there exists a proper filter $\mathcal{F}$ containing A with $\mathcal{F} \to x$.

Proof. First recall that $x \in \overline{A}$ if and only if every neighborhood of x nontrivially intersects A or equivalently if and only if the filterbase $\mathcal{B} = \{U \cap A\}_{U \in \mathcal{T}_x}$ does not contain the empty set. So if $x \in \overline{A}$, then simply generate a proper filter from $\mathcal{B}$. Conversely, if there is a proper filter $\mathcal{F}$ that converges to x and contains A, then $\mathcal{B} \subseteq \mathcal{F}$ and so $\mathcal{B}$ cannot contain the empty set. □

In the next theorem we'll show that filters also suffice to detect continuity. But first we need to understand functions in the context of filters.

Definition 3.6 Given a filter $\mathcal{F}$ on X and a function $f\colon X \to Y$, the set $\{fA \mid A \in \mathcal{F}\}$ of images of elements of $\mathcal{F}$ form a filterbase. The filter $f_*\mathcal{F}$ generated by this base is the *pushforward of $\mathcal{F}$ with respect to f*. Explicitly:

$$f_*\mathcal{F} := \{B \subseteq Y \mid \text{there exists } A \in \mathcal{F} \text{ with } fA \subseteq B\}$$

In this definition, the "generated by" is necessary since the set of images itself may not form a filter. For example, if f is not surjective, then the images don't contain Y and therefore cannot be upward closed.

Example 3.15 As a simple example, $f_*\mathcal{E}_{x_n} = \mathcal{E}_{fx_n}$. In other words, the pushforward of the eventuality filter of a sequence is the eventuality filter of the pushforward of that sequence.

Now here's the desired theorem.

Theorem 3.10 A function $f\colon X \to Y$ is continuous if and only if for every filter $\mathcal{F}$ on X, if $\mathcal{F} \to x$, then $f_*\mathcal{F} \to fx$.

Proof. Let $\mathcal{F}$ be a filter on X with $\mathcal{F} \to x$, and suppose $f\colon X \to Y$ is continuous. We want to show $\mathcal{T}_{fx} \subseteq f_*\mathcal{F}$; that is, for any $B \in \mathcal{T}_{fx}$ there exists a set $A \in \mathcal{F}$ with $fA \subseteq B$. So choose $A = f^{-1}B$. Continuity implies $f^{-1}\mathcal{T}_{fx} \subseteq \mathcal{T}_x$, which means $A \in \mathcal{T}_x$. The statement $\mathcal{F} \to x$ means $\mathcal{T}_x \subseteq \mathcal{F}$ and so $A \in \mathcal{F}$.

Conversely, suppose that whenever $\mathcal{F} \to x$ we have $f_*\mathcal{F} \to fx$ for any filter $\mathcal{F}$. Take $\mathcal{F}$ to be the filter generated by $\mathcal{T}_x$ to find that $f_*\mathcal{F} \to fx$, which means $\mathcal{T}_{fx} \subseteq f_*\mathcal{F}$. Thus for every $B \in \mathcal{T}_{fx}$, there exists a set A in $\mathcal{T}_x$ with $fA \subseteq B$. This proves that f is continuous. □

Filters, therefore, do indeed give the triad of theorems for all spaces.

all spaces (with filters)		*first countable spaces (with sequences)*
Theorem 3.8	**Hausdorff**	Theorem 3.5
Theorem 3.9	**closure**	Theorem 3.6
Theorem 3.10	**continuity**	Theorem 3.7

Having reached our goal, you might expect the chapter to conclude here. But not so fast. There's much more to filters. We've used them to study convergence, thereby promoting some "if, then" theorems about sequences to "if and only if" theorems about filters. But filters also shine particularly bright in discussions of compactness. To illustrate, we will use filters in the next section to provide a wonderfully short proof of Tychonoff's theorem, which was introduced—but not proven—in chapter 2.

3.4 Tychonoff's Theorem

The goal of this section is to prove the following theorem.

Tychonoff's Theorem 2 Given any collection $\{X_\alpha\}_{\alpha \in A}$ of compact spaces, the product $\prod_{\alpha \in A} X_\alpha$ is compact.

It's easier to prove that the product of finitely many compact spaces is compact than it is to prove the general case. For example, in Munkres' *Topology* (2000), compactness is introduced in chapter 3, where it is proven that the product of finitely many compact spaces is compact (Theorem 26.7). For the proof of the general case, the intrepid reader must wait until chapter 5 (Theorem 37.3), with a full chapter on countability and separation interrupting. Schaum's Outline (Lipschutz, 1965) states Tychonoff's theorem in chapter 12, but the proof is banished to the exercises. One must use the axiom of choice (or its equivalent) to prove the general case (see our theorem 3.14).

We will present a variation on Cartan's proof (1937a) by way of a little more filter technology. That technology is a particular kind of filter called an *ultrafilter*, which we introduce next. We'll take a leisurely stroll through the ideas, pointing out notable results along the way. In a grand finale, Tychonoff's theorem is proven in a few short lines in section 3.4.2.

3.4.1 Ultrafilters and Compactness

An *ultrafilter* is simply a filter that is *maximal*. The terms are synonymous.

Definition 3.7 A proper filter on a set is an *ultrafilter* if and only if it is not properly contained in any other proper filter.

This definition is second order since it deals with a quantification over subsets of a set. In practice, we'd like to work with a first-order definition—a characterization of an ultrafilter that doesn't require us to compare it to all other filters. Happily, such a characterization exists.

Proposition 3.1 A filter $\mathcal{U}$ on a set X is an ultrafilter if and only if for every subset $A \subseteq X$ the following condition holds: $A \notin \mathcal{U}$ if and only if there exists $B \in \mathcal{U}$ with $A \cap B = \varnothing$.

Proof. Let $\mathcal{U}$ be an ultrafilter. Then $A \notin \mathcal{U}$ if and only if the filter generated by $\mathcal{U} \cup \{A\}$ is the powerset 2^X. Since the generated filter consists of all sets containing an intersection of the form $B \cap A$ for some $B \in \mathcal{U}$, this is equivalent to the statement that the empty set contains $B \cap A$ for some $B \in \mathcal{U}$. Since the empty set is a subset of any set, the result follows.

Conversely, suppose $\mathcal{U}$ is a filter on X satisfying the condition, and let $\mathcal{F}$ be a filter properly containing $\mathcal{U}$. So there is at least one $A \in \mathcal{F}$ which is not in $\mathcal{U}$. By hypothesis, there must also exist a $B \in \mathcal{U}$ disjoint from A. But $\varnothing = A \cap B \in \mathcal{F}$, and so $\mathcal{F}$ is the improper filter by upward closure. □

Here's a nonexample followed by an example.

Example 3.16 If X has more than one point, then the *trivial filter* $\{X\}$ is not an ultrafilter.

Example 3.17 Given any x in a set X, the *principal filter at* x defined as $\{A \subseteq X \mid x \in A\}$ is an ultrafilter.

The next example highlights a point we wish to emphasize. In this chapter, we defined filters on powersets, but the definition of "downward directed, nonempty, upward closed" makes perfect sense in *any* poset. Keep this in mind. Filters in more general posets are useful, natural objects of study. This is illustrated well by *Riemann integration.*

Example 3.18 Let $f\colon [a,b] \to \mathbb{R}$ be a bounded function and let $(\mathcal{P}, \leq)$ be the poset of partitions of $[a,b]$ ordered by refinement. Any partition $P = \{a = x_0 < x_1 < \cdots < x_n = b\}$ yields two real numbers:

$$u_P := \textstyle\sum_{i=1}^n \sup\,(f|_{[x_{i-1},x_i]})(x_i - x_{i-1}) \qquad l_P := \textstyle\sum_{i=1}^n \inf\,(f|_{[x_{i-1},x_i]})(x_i - x_{i-1})$$

Similarly, any filter of partitions in $\mathcal{P}$ defines a pair of filters on $\mathbb{R}$:

$$\mathcal{B}_u(\mathcal{F}) = \{U \subseteq \mathbb{R} \mid \text{ there exists } Q \in \mathcal{F} \text{ such that } Q \leq P \text{ implies } u_P \in U\}$$

and

$$\mathcal{B}_l(\mathcal{F}) = \{U \subseteq \mathbb{R} \mid \text{ there exists } Q \in \mathcal{F} \text{ such that } Q \leq P \text{ implies } l_P \in U\}$$

For any ultrafilter $\mathcal{U}$ in $\mathcal{P}$, the filters $\mathcal{B}_u(\mathcal{U})$ and $\mathcal{B}_l(\mathcal{U})$ converge to real numbers. The function f is Riemann integrable if and only if they converge to the same value. In such a case, that value is called the integral $\int_a^b f$. So ultrafilters and the natural ordering of partitions allow us to replace the typical "partition norms" or "meshes" with a direct handling of the underlying orders. (As is often the case with categorical constructions, filters are reflective of what we have, not necessarily the computation of what we want.)

So just remember that a thorough discussion of filters can—and does—exist outside the context of powersets. But the powerset context is a very nice one: filters on powersets have special properties not shared by more general filters (see exercise 3.13 at the end of the chapter). In particular, because our filters are on powersets, the property of maximality is equivalent to another property—primality. In other words, ultrafilters on X are equivalent to *prime* filters on X. This reformulation will give us a simple characterization of compactness and ultimately a clean proof of Tychonoff's theorem.

Definition 3.8 A filter $\mathcal{F}$ on a set X is *prime* if and only if it is proper and if for all $A, B \subseteq X$,

$$A \cup B \in \mathcal{F} \text{ implies } A \in \mathcal{F} \text{ or } B \in \mathcal{F}$$

Theorem 3.11 A filter on X is maximal if and only if it is prime.

Proof. Suppose $\mathcal{F}$ is an ultrafilter on X and fails to be prime. Then there are $A, B \subseteq X$ such that $A \cup B \in \mathcal{F}$, but neither A nor B are in $\mathcal{F}$. The latter holds if and only if there exist sets $A', B' \in \mathcal{F}$ with $A \cap A' = \varnothing = B \cap B'$. This implies the intersection $(A \cup B) \cap (A' \cup B')$ is empty, which is true if and only if $A \cup B \notin \mathcal{F}$, a clear contradiction.

Now suppose $\mathcal{F}$ is prime but not a maximal. Then $\mathcal{F}$ is properly contained in a proper filter $\mathcal{G}$. So there exists a nonempty $A \in \mathcal{G}$ with $A \notin \mathcal{F}$. Notice that $X \setminus A \notin \mathcal{F}$, for otherwise $X \setminus A \in \mathcal{G}$ and if G contains both A and $X \setminus A$, then $\mathcal{G}$ would not be proper. But $A \cup (X \setminus A) = X \in \mathcal{F}$, contradicting the hypothesis that $\mathcal{F}$ is prime. □

Below, we'll use this theorem to give a succinct characterization of compact spaces. But despite the theorem, it will be good to keep a distinction between prime and maximal in our minds. There are several reasons why. First, as alluded to above, prime and maximal are not equivalent in more general settings. Keeping the two distinct in our minds strengthens our intuition. Second, a priori prime filters are difficult to construct. But as we'll soon see, we can always consider a maximal extension of a proper filter and then use the fact that ultrafilters are prime. Finally, the theorems proven below are naturally phrased in terms of prime filters because images commute with unions and therefore prime filters pushforward. Keeping the distinction thus makes proving theorems easier: if you need a prime filter, just construct one by extending a proper filter. If your construction involves pushing a filter forward, then prime filters are your friend. Now, en route to compactness let's prove our claim that every proper filter can be extended to a maximal one. We'll call on *Zorn's lemma*, whose statement we recount here.

Zorn's Lemma If every chain in a nonempty poset P has an upper bound, then P has a maximal element.

The Ultrafilter Lemma Every proper filter is contained in an ultrafilter.

Proof. Any set of filters $\{\mathcal{F}_\alpha\}_{\alpha \in A}$ is bounded above by the filter generated by finite intersections of elements of the $\{\mathcal{F}_\alpha\}$. When the set is a chain of proper filters, this upper bound is itself proper. Given a proper filter $\mathcal{F}$, chains of proper filters containing $\mathcal{F}$ thus have proper upper bounds. By Zorn's Lemma, there is a maximal filter containing $\mathcal{F}$. □

Corollary 3.11.1 Any infinite set has a non-principal ultrafilter.

Proof. Consider the Fréchet filter $\mathcal{F} := \{A \subseteq X \mid X \setminus A \text{ is finite}\}$ and appeal to the Ultrafilter Lemma to extend $\mathcal{F}$ to an ultrafilter $\mathcal{U}$. Were $\mathcal{U}$ to contain any finite set, it would contain its (cofinite) complement; hence $\varnothing \in \mathcal{U}$, which contradicts that $\mathcal{U}$ is a proper filter. □

To appreciate this result, think back to example 3.17. When pressed, it's hard to come up with ultrafilters that are not principal. The fact that any *nonprincipal ultrafilters* exist is not at all obvious. To produce one, we needed to appeal to the Ultrafilter Lemma. The Ultrafilter Lemma also gives the promised characterization of compactness.

Theorem 3.12 A space X is compact if and only if every prime filter converges.

Proof. Suppose $\mathcal{F}$ is a prime filter that fails to converge to any $x \in X$. Equivalently, suppose for all x there exists $U_x \in \mathcal{T}_x - \mathcal{F}$. The set $\{U_x\}_{x\in X}$ is an open cover. By compactness, choose a finite subcover $\{U_{x_i}\}_{i=1}^n$. Then $U_{x_1} \cup \cdots \cup U_{x_n} = X \in \mathcal{F}$. By primality, there exists an i such that $U_{x_i} \in \mathcal{F}$, a contradiction.

Now suppose X is not compact. Choose a collection $\mathcal{V}$ of closed sets with the finite intersection property and empty intersection. Note that for all x there exists $V_x \in \mathcal{V}$ with $x \notin V_x$. Further, by the Ultrafilter Lemma, $\mathcal{V}$ is contained in an ultrafilter $\mathcal{U}$. However, $\mathcal{U} \not\to x$ for any x, for otherwise it would imply $\varnothing = V_x \cap V_x^c \in \mathcal{U}$ contradicting properness of $\mathcal{U}$. □

You'll recognize this theorem as a generalization of the Bolzano-Weierstrass theorem introduced in chapter 2. Restated, it says:

If X is compact, then every sequence has a convergent subsequence.

So recalling from example 3.14 that subsequences correspond to larger (eventuality) filters and that filters are good at promoting "if then" theorems to "if and only if" ones, you might hope for a promotion of Bolzano-Weierstrass for filters:

X is compact if and only if every proper filter is contained in a convergent proper filter.

As a result of theorem 3.12, this is indeed the case. Where we once used convergent subsequences, we now use prime filters. Another consequence of theorem 3.12 comes for free when paired with theorem 3.8.

Corollary 3.12.1 A space X is compact Hausdorff if and only if every prime filter has exactly one limit point.

This characterization of compact Hausdorff spaces is the beginning of a long categorical tale. Sharing the full story would take us too far off course, so instead we'll tell an abridged version. It starts with the fact that ultrafilters define a functor from the category Set to itself, a consequence of this next theorem.

Theorem 3.13 Let $\mathcal{U}$ be an ultrafilter on X and let $f\colon X \to Y$. The pushforward $f_*\mathcal{U}$ is an ultrafilter on Y.

Proof. Exercise. □

Since the pushforward of an ultrafilter is an ultrafilter, the assignment $\beta\colon \mathsf{Set} \to \mathsf{Set}$ that sends a set X to βX, the set of ultrafilters on X, defines a functor. For a morphism $f\colon X \to Y$ of sets, the function $\beta f\colon \beta X \to \beta Y$ sends an ultrafilter to its pushforward. In the special case when X is a compact Hausdorff space, every ultrafilter on X converges to exactly one point. So you might wonder whether the assignment $\alpha\colon \beta X \to X$ that sends an ultrafilter to its unique limit point is of any interest. It is. It's the key to unlocking an important categorical statement:

> The category of compact Hausdorff spaces is equivalent to the category of algebras for the ultrafilter monad.

What's the ultrafilter monad? And what does it mean to be an algebra for one? We won't get into the details, but we will try to give you an idea about what the statement means. Principal filters play a principal role: since the pushforward of a principal filter is principal, they assemble into a natural transformation $\eta : \mathrm{id}_{\mathsf{Set}} \to \beta$ defined by

$$\eta_X(x) = P_x$$

where P_x is the principal filter at $x \in X$. There is another natural transformation $\mu\colon \beta \circ \beta \to \beta$ that comes into play. We will not describe μ except to say that it works like a kind of multiplication, and the natural transformation η behaves like a unit for this multiplication. The triple (β, η, μ) defines something called a *monad.* (It should remind you of a *monoid*, which also consists of three things: a set X, an associative multiplication map $m\colon X \times X \to X$, and an element $u\colon * \to X$ behaving as a unit for m.) Once you have a monad, you can define something called an algebra for that monad, and the algebras for a monad form a category. And *this* category, one can show, is equivalent to the category of compact Hausdorff spaces. Behind the curtain of this categorical connection between ultrafilters and compact Hausdorff spaces is the rich theory of *adjunctions*. For this reason, the story will naturally resurface later in this book: it's closely related to the discussion of the Stone-Čech compactification in chapter 5. For an introduction to monads, see Riehl (2016), and for a fuller account of the ultrafilter monad tale, see Manes (1969) and the article on compacta at the nLab (Stacey et al., 2019). The ambitious reader may further enjoy Leinster (2013).

After this leisurely stroll through ultrafilters, prime filters, Bolzano-Weierstrass, and monads, we are now ready to prove that which we set out to prove: Tychonoff's theorem.

3.4.2 A Proof of Tychonoff's Theorem

There is a conservation of difficulty in mathematics. One theorem may have many proofs, and more sophisticated tools will give more elegant proofs. Historically, the difficulties in Tychonoff's theorem were in finding the correct definition of the product topology and in characterizing compactness. By using ultrafilters to characterize compactness, we are using a sophisticated theoretical tool. The proof we share is correspondingly elegant (Chernoff, 1992).

If sequences had been sufficient for discussions of convergence, then we could use a ready-made argument. Just use continuity of projection maps from the product to push a sequence forward. In each factor, pass to a subsequence, which converges by Bolzano-

Weierstrass. Then in the product, form a subsequence by taking the indices common to all the convergent subsequences in the projections, and conclude this sequence converges. Here's the thrill: by replacing sequences with filters and doing the work needed to develop some understanding of the theory, this ready-made argument becomes a genuine proof.

Proof of Tychonoff's theorem. Let $\{X_\alpha\}_{\alpha\in A}$ be a family of compact spaces, define $X := \prod_{\alpha\in A} X_\alpha$, and let $\mathcal{F}$ be an ultrafilter on X. We must show that $\mathcal{F}$ converges.

The pushforward of an ultrafilter is an ultrafilter and, since the X_α are compact, there exists x_α such that $(\pi_\alpha)_*\mathcal{F} \to x_\alpha$. So by definition, for all open neighborhoods U of x_α, there exists $B \in \mathcal{F}$ with $\pi_\alpha B \subseteq U$. Equivalently, $B \subseteq \pi_\alpha^{-1}U$, and so $\pi_\alpha^{-1}U \in \mathcal{F}$. And every open neighborhood of $(x_\alpha)_{\alpha\in A} \in X$ is a union of finite intersections of the $\pi_\alpha^{-1}U$. Therefore, $\mathcal{F} \to (x_\alpha)_{\alpha\in A}$. □

We'll close this chapter with the remark that any treatment of Tychonoff's theorem requires some machinery from set theory. In our presentation, we hid the machinery in Zorn's lemma, which we used to prove the Ultrafilter Lemma. The reason set theory is unavoidable is because Tychonoff's theorem is equivalent to the axiom of choice. Without digressing too much, we'd like to give you some idea of why *Zermelo-Frankel-Choice* is equivalent to *Zermelo-Frankel-Tychonoff.*

3.4.3 A Little Set Theory

In any imaginable proof that the Tychonoff theorem implies the axiom of choice, one begins with an arbitrary collection of sets and then creates a collection of compact topological spaces. The compactness of the product leads to the existence of a choice function. In 1950, Kelley proved that the Tychonoff theorem implies the axiom of choice (Kelley, 1950) using augmented cofinite topologies. Here, we give an easier proof. First, we recall the axiom of choice.

The Axiom of Choice For any collection of nonempty sets $\{X_\alpha\}_{\alpha\in A}$, the product $\prod_{\alpha\in A} X_\alpha$ is nonempty.

Theorem 3.14 Tychonoff's theorem is equivalent to the axiom of choice.

Proof. We used Zorn's lemma to prove Tychonoff's theorem. Although we don't prove it, the axiom of choice implies Zorn's lemma (see exercise 3.14 at the end of the chapter), from which it follows that Tychonoff's theorem is implied by the axiom of choice.

To prove that Tychonoff's theorem implies the axiom of choice, let $\{X_\alpha\}_{\alpha\in A}$ be a collection of nonempty sets. We need to make a bunch of compact spaces so we can apply the Tychonoff theorem. First, add a new element to X_α called "∞_α," letting $Y_\alpha = X_\alpha \cup \{\infty_\alpha\}$. Each set Y_α can be made into a space by defining the topology to be $\{\varnothing, \{\infty_\alpha\}, X_\alpha, Y_\alpha\}$. Note that Y_α is compact—there are only finitely many open sets so every open cover is finite. Thus by Tychonoff's theorem, $Y := \prod_{\alpha\in A} Y_\alpha$ is compact.

Now consider a collection of open sets $\{U_\beta\}_{\beta\in A}$ of Y where U_β is the basic open set in Y obtained by taking the product of all Y_αs for $\alpha \neq \beta$ and putting the open set $\{\infty_\beta\}$ in the βth factor. Notice that any finite subcollection $\{U_{\beta_1}, \ldots, U_{\beta_n}\}$ cannot cover Y, for the function f defined as follows is not in $\cup_{i=1}^{n} U_{\beta_i}$. Choose a partial function $\overline{f} \in \prod_{i=1}^{n} X_{\beta_i}$, which is possible without the axiom of choice since the product is finite. Then extend $\overline{f}$ to a function $f \in Y$ by setting $f(\alpha) = \infty_\alpha$ for all $\alpha \neq \beta_1, \ldots, \beta_n$, which is possible since we're not making any choices.

Therefore, the collection $\{U_\beta\}$ cannot cover Y. So there is a function $f \in Y$ not in the $\cup_{\alpha\in A} U_\alpha$. This says that for no $\alpha \in A$ does $f\alpha = \infty_\alpha$. Therefore, $f\alpha \in X_\alpha$ for each α, which is a desired choice function. □

Exercises

1. Suppose A is a subspace of X. We say a map $f\colon A \to Y$ can be *extended* to X if there is a continuous map $g\colon X \to Y$ with $g = f$ on A.

   a) Prove that if A is dense in X and Y is Hausdorff, then f can be extended to X in at most one way.

   b) Give an example of spaces X and Y, a dense subset A, and a map $f\colon A \to Y$ such that f can be extended to X in more than one way.

   c) Give an example of spaces X and Y, a dense subset A, and a map $f\colon A \to Y$ such that f cannot be extended.

2. Prove that $\mathbb{R}$ with the cocountable topology (sets with countable complement are open) is a non-Hausdorff space in which convergent sequences have unique limits.

3. Check all the details of example 3.14.

4. Check all the details of example 3.5.

5. Nets are an earlier generalization of sequences introduced by Moore and Smith (1922); Moore (1915) used to address the insufficiency of sequences. This exercise demonstrates that nets and proper filters are formally interchangeable.

   Definition 3.9 A *net* is a function $\eta\colon S \to X$ whose domain is a directed set.
   A *directed set* is defined to be a pair $(S, \leq)$, where S is a set and $\leq$ is a relation on S satisfying:

   - for all $s \in S$, $s \leq s$,
   - for all $s, t, u \in S$, $s \leq t$ and $t \leq u$ imply $s \leq u$,
   - for all $s, t \in S$, there exists $u \in S$ with $s \leq u$ and $t \leq u$.

   We say that η *converges* to x if and only if its eventuality filter

   $$\mathcal{E}_\eta := \{A \subseteq X \mid \text{ there exists } t \in S \text{ such that } s \geq t \text{ implies } \eta s \in A\}$$

   contains $\mathcal{T}_x$ in which case we write $\eta \to x$.

   a) A sequence is an example of a net. Show that a subsequence of a sequence is a subnet, but not all subnets of a sequence are subsequences. For an interesting example, use the family of sawtooth functions from example 3.5 whose corners have rational coordinates.

   b) Given a proper filter $\mathcal{F}$, let $\mathcal{D} := \{(A, a) \in 2^X \times X \mid a \in A \in \mathcal{F}\}$. Show that $\mathcal{D}$ is directed by the relation $(A, a) \leq (B, b)$ if and only if $B \subseteq A$.

   c) Let $\pi_\mathcal{F}\colon \mathcal{D} \to X$ be the net sending $(A, a) \mapsto a$. Prove that $\mathcal{E}_{\pi_\mathcal{F}} = \mathcal{F}$.

   d) Conclude that $\pi_\mathcal{F} \to x$ if and only if $\mathcal{F} \to x$.

6. Check all the details of example 3.6.

7. **Pushforward of Filters:** Given $f\colon X \to Y$ and a filter $\mathcal{F}$ on X, prove that the set

$$\{B \subseteq Y \mid \text{ there exists } A \in \mathcal{F} \text{ such that } fA \subseteq B\}$$

is a filterbase.

8. Prove theorem 3.13.

9. Here are two variations of Hausdorff. Call a space KC if all its compact sets are closed. Call a space US if the limits of convergent sequences are unique. Prove that Hausdorff implies KC implies US, but that the implications are strict (Wilansky, 1967).

10. Show that a countable intersection of open dense sets in a complete metric space is dense. (This is called the *Baire category theorem.*)

11. Let X be a compact space and let $\{f_n\}$ be an increasing sequence in $\mathsf{Top}(X, \mathbb{R})$. Prove that if $\{f_n\}$ converges pointwise, then $\{f_n\}$ converges uniformly.

12. Verify that the following definition of filters in posets specializes to the definition we gave for filters on sets.

Definition 3.10 A *filter in a poset* $(\mathcal{P}, \preceq)$ is a collection $\mathcal{F} \subseteq \mathcal{P}$ such that it is:

Downward directed: $a, b \in \mathcal{F}$ implies there exists $c \in \mathcal{P}$ such that $c \preceq a$ and $c \preceq b$,

Nonempty: $\mathcal{F}$ is nonempty,

Upward closed: $a \in \mathcal{F}$ and $a \preceq b$ implies $b \in \mathcal{F}$.

13. Following up on our discussion of prime and maximal filters, consider the following pair of lattices:

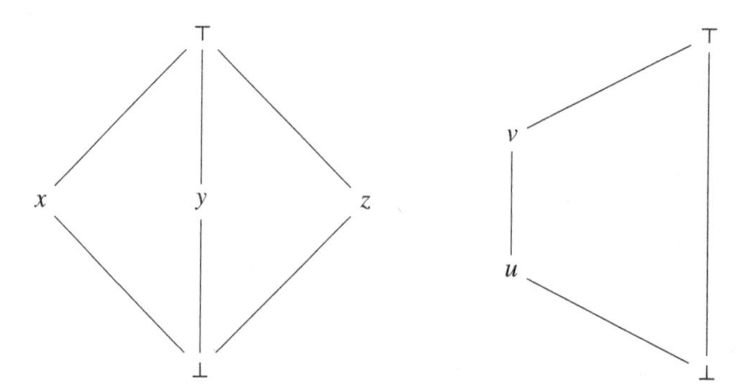

Find a maximal filter that is not prime in the lattice on the right and a filter that is prime but not maximal in the lattice on the left.

Note: Lattices without any sublattice isomorphic to one of these satisfy the distributive property $x \wedge (y \vee z) = (x \wedge y) \vee (x \wedge z)$. And in particular, we may conclude that for distributive lattices, maximal filters are prime by recycling our proof substituting unions for joins and intersections for meets.

14. Transfinite Induction: Axiom of choice implies Zorn's lemma

Ordinals are set theory's response to the question of how things may be ordered "one after another." Consequently, they form the setting in which induction may be defined.

Definition 3.11 A *well ordering* on a set S is a linear (or total) order "$\leq$" in which every nonempty subset has a least element. Together with order preserving—that is, monotone—functions, well-orders form a category. *Ordinals* are defined to be the isomorphism classes of objects in this category. Following von Neumann, we associate to each ordinal $[\alpha]$ a representative well-ordering:

$$\alpha := \{ \text{ ordinals } \beta < \alpha \}$$

and will feel free to refer to "the" ordinal α as this representative well-order.

Being well-ordered amounts to having two pieces of information: there's always a starting point and at every element there is an unambiguously defined next element—precisely the information needed to carry out induction. It's worth looking at the first few familiar ordinals to get a feel for them:

Example 3.19 First observe that there is a least ordinal, typically called 0, namely the initial well-order $\varnothing$ with the empty relation. This is the seed from which we may freely generate ordinals.

Names	*Representatives*	*Orders*
0	$\varnothing$	
1	$\{0\}$	0
2	$\{0, 1\}$	$0 \to 1$
$\vdots$	$\vdots$	$\vdots$
ω	$\mathbb{N}$	$0 \to 1 \to \cdots$
$\omega + 1$		$0 \to 1 \to \cdots \omega$
$\omega + 2$		$0 \to 1 \to \cdots \omega \to \omega + 1$
$\vdots$	$\vdots$	$\vdots$

Some caution is in order. Note that the underlying sets of, say, ω and $\omega + 1$ are in bijection, but as orders they are distinct. For example, the ordinal $\omega + 1$ has a nonzero element that is not the immediate successor of any other element. Such elements are called *limit ordinals* and play an important role in the theory of ordinals. Second, keep in mind that not every total order is a well-order; the rationals are not well ordered, for example. Finally, it is tempting to guess that the ordinals are themselves well ordered. After all, they are linearly ordered by "is an initial segment of," and in any set of ordinals, there is a least element. However, this leads to the *Burali-Forti paradox*: were Ω the *set* of all ordinals, then it's well ordered by our above comments and hence is (order isomorphic to) an ordinal. So $\Omega < \Omega$, contradicting trichotomy—we are forced to conclude that the collection of all ordinals is not a set.

(i) Use the Burali-Forti paradox to prove that there are ordinals of arbitrarily large cardinality. (Hint: try contradiction.)

(ii) Demonstrate the following: to prove a property $P(-)$ holds for all ordinals, it is sufficient to demonstrate that:

- *Base Case*: $P(0)$ holds.
- *Successor Step*: $P(\alpha)$ implies $P(\alpha + 1)$.
- *Limit Step*: Given limit ordinal λ, for all $\alpha < \lambda$ such that $P(\alpha)$ holds, then $P(\lambda)$ holds.

(iii) Using the following strategy, prove that the axiom of choice implies Zorn's lemma:

Given the axiom of choice and a nonempty partially ordered set $(\mathcal{P}, \leq)$ in which every chain has an upper bound, for each $a \in \mathcal{P}$ define

$$E_a := \{b \in \mathcal{P} \mid a < b\}$$

There are two cases: if there is some a for which $E_a = \varnothing$, then a is maximal and we are done; otherwise, the axiom of choice guarantees that there is a function $f \colon \mathcal{P} \to \mathcal{P}$ such that $fa \in E_a$ for all a.

Use transfinite induction to prove that there are chains of all ordinal lengths in $\mathcal{P}$. State a contradiction and conclude that the axiom of choice implies Zorn's lemma.

4 Categorical Limits and Colimits

A comathematician is a device for turning cotheorems into ffee.
—Unknown (Please tell us if you made this joke up!)

Introduction. Categorical limits and colimits are one of the—and in some sense *the most*—efficient way to build a new mathematical object from old objects. The constructions introduced in chapter 1—subspaces, quotients, products, and coproducts—are examples in Top, though the discussion can occur in any category. In fact, there are a number of other important constructions—pushouts, pullbacks, direct limits, and so on—so it's valuable to learn the general notion.

In practice, limits are typically built by picking a subcollection according to some constraint, whereas colimits are typically built by "gluing" objects together. More formally, the defining property of a limit is characterized by morphisms whose *domain* is the limit. The defining property of a colimit is characterized by morphisms whose *codomain* is the colimit. Because of their generality, limits and colimits appear all across the mathematical landscape. A direct sum of abelian groups, the least upper bound of a poset, and a CW complex are all examples of limits or colimits, and we'll see more examples in the pages to come. Section 4.1 opens the chapter by answering the anticipated question, "A (co)limit of *what*?" The remaining two sections contain the formal definition of (co)limits followed by a showcase of examples.

4.1 Diagrams Are Functors

In topology, one asks for the limit of a sequence. In category theory, one asks for the (co)limit of a *diagram*. In what follows, it will be helpful to view a diagram as a functor. More specifically, a diagram in a category is a functor from the shape of the diagram to the category. For example, a commutative diagram like this

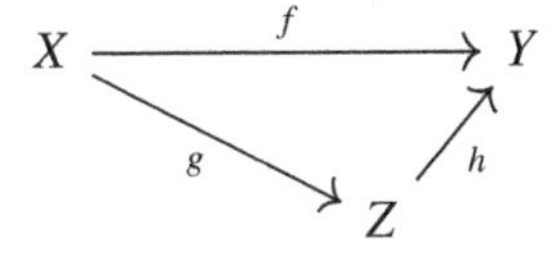

in a category C is a choice of three objects X, Y, and Z and some morphisms $f\colon X \to Y$, $g\colon X \to Z$, and $h\colon Z \to Y$, with $f = hg$. It can be viewed as the image of a functor from an *indexing category*; that is, from a picture like this:

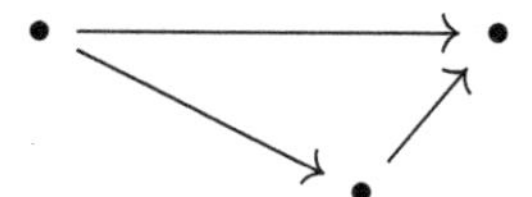

This is a small category—let's call it D—containing three objects pictured as bullets and three morphisms pictured as arrows. Though they must be in D, not shown are the identity morphisms and compositions. Here composition is determined by setting the composition of the two diagonal arrows to be the horizontal arrow. A functor $F\colon \mathsf{D} \to \mathsf{C}$ involves a choice of three objects and three morphisms and must respect composition. In summary,

$$\text{a diagram} \left(\begin{array}{ccc} X & \xrightarrow{f} & Y \\ & {\scriptstyle g}\searrow \quad \nearrow{\scriptstyle h} & \\ & Z & \end{array} \in \mathsf{C} \right) \quad \text{is a functor} \left(\begin{array}{ccc} \bullet & \longrightarrow & \bullet \\ & \searrow \quad \nearrow & \\ & \bullet & \end{array} \to \mathsf{C} \right)$$

The concept of identifying a map with its image is a familiar one. A sequence of real numbers, for instance, is a function $x\colon \mathbb{N} \to \mathbb{R}$, though one may write x_n for xn and think of the sequence as the collection $(x_1, x_2, \ldots)$. Likewise, a path in a topological space X is a continuous function $p\colon [0, 1] \to X$, though one may often have the image $pI \subset X$ in mind. The idea that "a diagram is a functor" is no different.

Definition 4.1 Let D be a small category. A D-*shaped diagram* in a category C is a functor $\mathsf{D} \to \mathsf{C}$. If the categories C and D are understood, we'll simply say *diagram* instead of D-shaped diagram in C.

Because a diagram is a functor, it makes sense to ask for a morphism from one diagram to another—it's a natural transformation of functors. As we'll see below, a (co)limit of a diagram F involves a morphism between F and a diagram of a specific shape—a point. A point-shaped diagram is a functor that is constant at a given object of a category. Indeed we can view any object A of C as a D-shaped diagram for any category D and namely as the *constant functor*. It is defined by sending every object in D to A and every arrow in D to the identity at A in C.

$$\begin{array}{ccc} \bullet & & A \\ \downarrow & \overset{A}{\longmapsto} & \downarrow{\scriptstyle \mathrm{id}_A} \\ \bullet & & A \end{array}$$

Notice we're using the symbol A for both the object A and for the constant functor itself. In other words, we allow ourselves flexibility in viewing A as an object or as a functor. In this way, we introduce the phrase "a map from an object to a diagram" to mean a natural transformation from the constant functor to the diagram.

Definition 4.2 Given a functor $F\colon \mathsf{D} \to \mathsf{C}$, a map from an object A to F—that is, an element of $\mathsf{Nat}(A, F)$—is called a *cone from A to F*. Similarly, a *cone from F to A* is an element of $\mathsf{Nat}(F, A)$.

Unwinding the definition, a cone from A to $F\colon \mathsf{D} \to \mathsf{C}$ is a collection of maps

$$\{\, A \xrightarrow{\eta_\bullet} F\bullet \quad \text{where} \quad \bullet \quad \text{is an object in } \mathsf{D} \,\}$$

such that the diagrams:

$$\begin{array}{ccc} & A & \\ {\scriptstyle \eta_\bullet}\swarrow & & \searrow{\scriptstyle \eta_\circ} \\ F\bullet & \xrightarrow{\;F\varphi\;} & F\circ \end{array}$$

commute for every morphism $\bullet \xrightarrow{\varphi} \circ$ in D. For example, a cone from some object A to the functor F with which we opened the section consists of three maps η_X, η_Y, and η_Z fitting together in a commutative diagram:

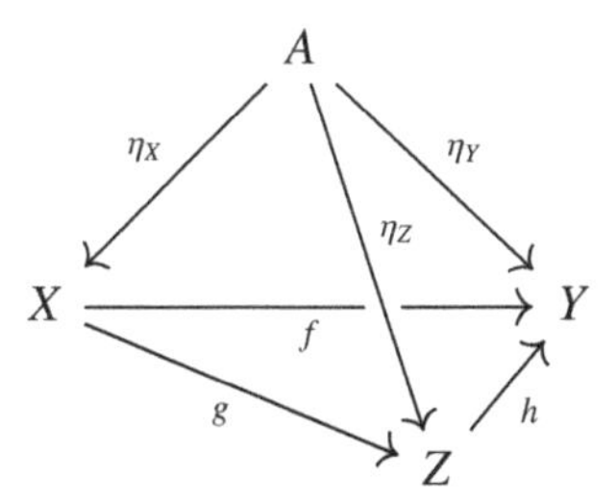

As we'll see in the next section, a limit of F is a special cone over F and a colimit of F is a special cone under F.

4.2 Limits and Colimits

Here are the formal definitions of limit and colimit.

Definition 4.3 A *limit* of the diagram $F\colon \mathsf{D} \to \mathsf{C}$ is a cone η from an object $\lim F$ to the diagram satisfying the universal property that for any other cone γ from an object B to the diagram, there is a unique morphism $h\colon B \to \lim F$ so that $\gamma_\bullet = \eta_\bullet h$ for all objects $\bullet$ in D.

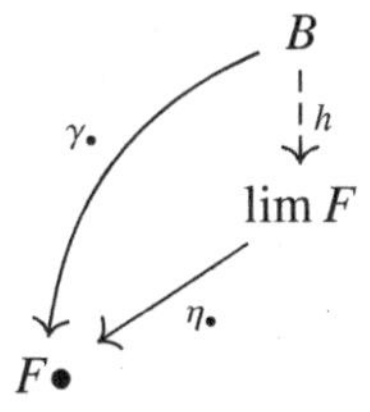

One may understand this colloquially in the following way. First, recognize that there may exist *many* cones over F—many objects with maps pointing down to the diagram. But only *one* of them can satisfy the property of limit, namely $\eta\colon \lim F \to F$. You might, however, come across another cone $\gamma\colon B \to F$ that behaves similarly. Perhaps γ also commutes with every arrow in the diagram F and thus seems to imitate η. But the similarity is no coincidence. The natural transformation γ behaves like η precisely because *it is built up from* η. That is, it factors through η as $\gamma = \eta \circ h$ for some unique morphism h. This is the universality of the limit cone. Informally, then, the limit of a diagram is the "shallowest" cone over the diagram. One might visualize all possible cones over the diagram as cascading down to the limit. It is the one that is as close to the diagram as possible:

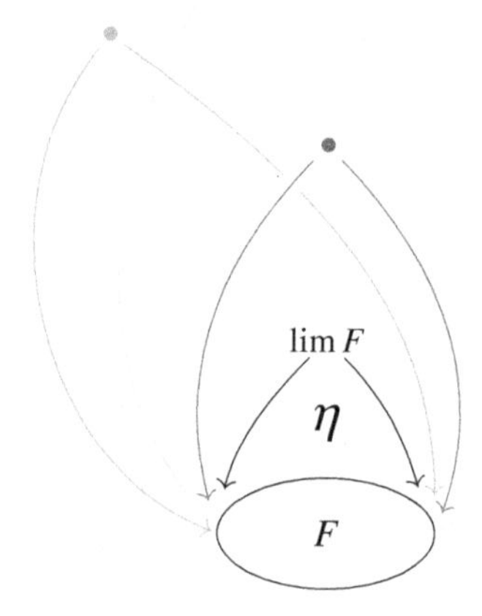

We can similarly ask for maps from a diagram to an object in the category. This leads to the following definition.

Definition 4.4 A *colimit* of the diagram $F\colon \mathsf{D} \to \mathsf{C}$ is a cone ϵ from the diagram to an object colim F satisfying the universal property that for any other cone γ from the diagram to an object B, there is a unique map $h\colon \operatorname{colim} F \to B$ so that $\gamma_\bullet = h\epsilon_\bullet$ for all objects $\bullet$ in D.

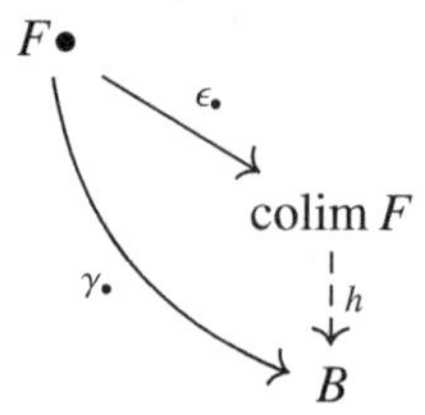

Informally, a colimit of a diagram F is the "shallowest" cone under F. There many exist *many* cones—many objects with maps pointing away from the diagram—under F, but a colimit is the cone that is closest to the diagram. Again, one makes sense of the informal words "shallow" and "close" via the universal property in the definition:

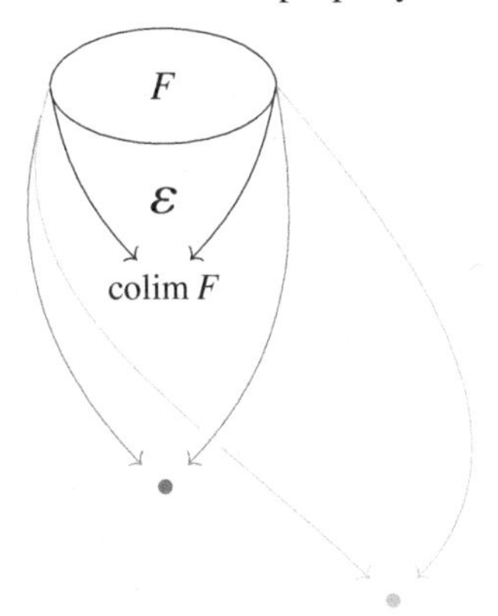

Be aware that the (co)limit of a diagram may not exist, but if it does, then—as the reader should verify—it is unique up to a unique isomorphism. We will therefore refer to *the* (co)limit of a diagram.

4.3 Examples

Depending on the shape of the indexing category, the (co)limit of a diagram may be given a familiar name: intersection, union, Cartesian product, kernel, direct sum, quotient, fibered product, and so on. The following examples illustrate this idea. In each case, recall that the data of a (co)limit is an object together with maps to or from that object, satisfying a universal property.

4.3.1 Terminal and Initial Objects

If the indexing category D is empty—no objects and no morphisms—then a functor $\mathsf{D} \to \mathsf{C}$ is an empty diagram. The limit of an empty diagram is called a *terminal object*. It is an object T in C such that for every object X in C there is a unique morphism $X \to T$. In other words, all objects in the category terminate at T. In Set the terminal object is the one-point set; in Top it's the one-point space; in Grp it's the trivial group; in FVect, it's the zero vector space; in a poset, its the greatest element, if it exists. This highlights an important point: not every category has a terminal object. For example, $\mathbb{R}$ with the usual ordering is a poset without a greatest element—it is a category without a terminal object.

Dually, the colimit of an empty diagram is called an *initial object*. It is an object I in C such that for every object X in C there is a unique morphism $I \to X$. In other words, all objects in the category initialize from I. In Set the initial object is the empty set; in Top

it's the empty space; in Grp it's the trivial group; in FVect, it's the zero vector space; in a poset, it's the least element, if it exists. Again, not every category has an initial object.

Many times, when one is interested in the limit of a diagram, the colimit of the diagram will be trivial, or vice versa. For example, the colimit of a diagram that has a terminal object Y is just the object Y, together with the morphisms in the diagram. For example, the object Y is terminal in this diagram:

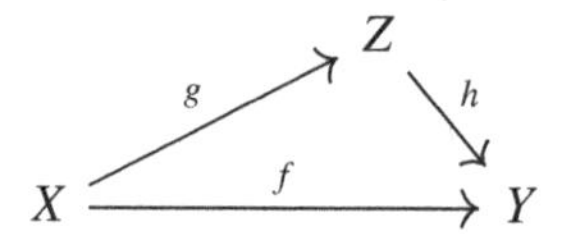

and indeed the colimit of this diagram is just Y with the map of the diagram given by $f: X \to Y$, $h: Z \to Y$, and $\mathrm{id}_Y: Y \to Y$. It has the universal property since for any object S, a map from the diagram to S includes a map from Y to S making everything commute. That map is the one satisfied by the universal property of Y being the colimit.

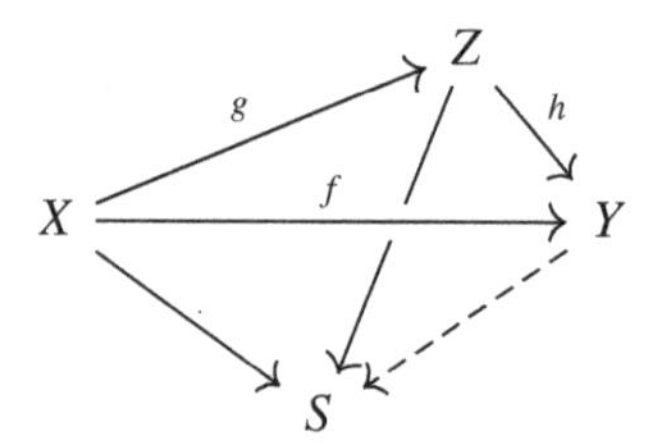

Similarly, the limit of a diagram with an initial object X is just the initial object X, together with the morphisms in the diagram.

4.3.2 Products and Coproducts

If the indexing category D has no nonidentity morphisms—that is, if D is a *discrete category*—then a diagram $\mathsf{D} \to \mathsf{C}$ is just a collection of objects parametrized by D. In this case, the limit of the diagram is called the *product* and the colimit is called the *coproduct*. When $\mathsf{C} = \mathsf{Set}$, you can verify the universal properties of the product and coproduct given in chapter 0 and show that they are indeed the limit and colimit of discrete diagrams. When the category is Top, the limit is the product of the spaces in the diagram, equipped with the product topology, together with projection maps, down to each of the factors. Likewise, the coproduct is the disjoint union of spaces, equipped with the coproduct topology, together with inclusion maps from each space.

Quite often, one is only interested in a subset (or subspace) of the product. For instance, given sets or spaces X and Y, it's often interesting to consider only those pairs $(x, y) \in X \times Y$ where x and y relate to each other according to some equation. In a dual sense, one may be interested in identifying parts of sets (or spaces) rather than considering their full coproduct. The next few categorical constructions provide different ways of accomplishing these tasks.

4.3.3 Pullbacks and Pushouts

A functor from $\bullet \to \bullet \leftarrow \bullet$ is a diagram

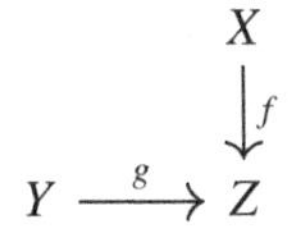

$$\begin{array}{ccc} & & X \\ & & \downarrow f \\ Y & \xrightarrow{g} & Z \end{array}$$

and its limit is called the *pullback* of X and Y along the morphisms f and g. In **Set** the pullback is realized by the set consisting of all pairs (x, y) satisfying $fx = gy$, along with projection maps onto each factor X and Y. The set is denoted by $X \times_Z Y$. Diagrammatically, there is a special notation to describe pullbacks. A square diagram decorated with a caret "$\lrcorner$" in the upper left corner denotes a pullback diagram. For example, this diagram

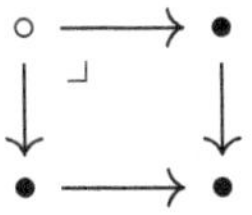

$$\begin{array}{ccc} \circ & \longrightarrow & \bullet \\ \downarrow & \lrcorner & \downarrow \\ \bullet & \longrightarrow & \bullet \end{array}$$

should be read as saying, "the square commutes and the object $\circ$ is the pullback." As a concrete example, suppose $X = *$ is the one-point set so that a function $f: * \to Z$ picks out an element $z \in Z$. Then the pullback consists of the set of points $y \in Y$ such that $gy = z$. In other words, the pullback is the preimage $g^{-1}z \subset Y$.

In **Top**, the pullback has $X \times_Z Y$ as its underlying set and it becomes a topological space when viewed as a subspace of the product $X \times Y$. It satisfies the universal property described by this diagram:

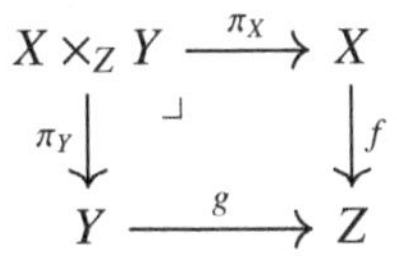

$$\begin{array}{ccc} X \times_Z Y & \xrightarrow{\pi_X} & X \\ \pi_Y \downarrow & \lrcorner & \downarrow f \\ Y & \xrightarrow{g} & Z \end{array}$$

Explicitly, the pullback topology (first characterization) is the finest topology for which the projection maps $\pi_x: X \times_Z Y \to X$ and $\pi_Y: X \times_Z Y \to Y$ are continuous. Alternatively, the pullback topology (second characterization) is determined by specifying that maps into the pullback from any space W are continuous if and only if the maps $W \to X$ and $W \to Y$ obtained by postcomposing with f and g are continuous. Said the other way around: maps to the pullback from a space W are specified by maps $a: W \to X$ and $b: W \to Y$ with $fa = gb$.

At some point you might encounter a statement such as "the map $p: Y \to X$ is the *pullback of* $\pi: E \to B$ *along the map* $f: X \to B$." This means that p fits into a pullback square with f and π as pictured

$$\begin{array}{ccc} Y & \longrightarrow & E \\ {\scriptstyle p}\downarrow & \lrcorner & \downarrow{\scriptstyle \pi} \\ X & \xrightarrow{f} & B \end{array}$$

and that Y together with $p\colon Y \to X$ and the unnamed map is the pullback of the rest of the diagram. The unnamed map $Y \to E$ exists and is part of the pullback but might not be explicitly mentioned.

Dually, a functor from $\bullet \leftarrow \bullet \rightarrow \bullet$ is a diagram,

$$\begin{array}{ccc} Z & \xrightarrow{g} & X \\ {\scriptstyle f}\downarrow & & \\ Y & & \end{array}$$

and its colimit is called the *pushout* of X and Y along the morphisms f and g. In Set the pushout of maps $f\colon Z \to X$ and $g\colon Z \to Y$ is realized by the quotient set $X \sqcup_Z Y := X \coprod Y/\sim$ where $\sim$ is the equivalence relation generated by $fz \sim gz$ for all $z \in Z$, together with maps from X and Y into the quotient. By analogy with pullbacks, we will use a caret to denote a pushout square. That is, a diagram such as

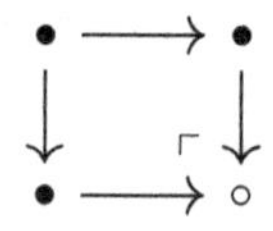

means that the square commutes and that $\circ$ is the pushout. As a concrete example of a pushout in Set, note that given any two sets A and B, the intersection $A \cap B$ is a subset of A and B. So we have the diagram of inclusions

$$\begin{array}{ccc} A \cap B & \hookrightarrow & A \\ \downarrow & & \\ B & & \end{array}$$

The pushout of this diagram is the union $A \cup B$. Said explicitly, the union fits into the diagram, and for any other set S that fits into the diagram, there's a unique function from $A \cup B \to S$ as pictured by the dashed line below:

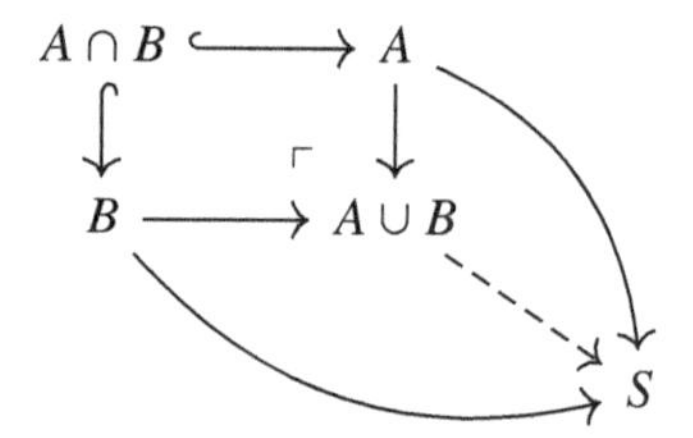

The universal property of the pushout in this example says that a function $A \cup B \to S$ is the same as a pair of functions $A \to S$ and $B \to S$ that agree on $A \cap B$.

In Top, the pullback has the quotient $X \sqcup_Z Y$ as its underlying set. It becomes a topological space as a quotient space of the coproduct. This pushout satisfies the universal property described by this diagram; that is, by making use of the caret notation for pushouts:

$$\begin{array}{ccc} Z & \xrightarrow{f} & X \\ {\scriptstyle g}\downarrow & \ulcorner & \downarrow{\scriptstyle i_X} \\ Y & \xrightarrow{i_Y} & X \sqcup_Z Y \end{array}$$

The pushout topology (first characterization) is the coarsest topology for which the maps $i_X : X \to X \sqcup_Z Y$ and $i_Y : Y \to X \sqcup_Z Y$, which send an element to its equivalence class, are continuous. Alternatively, the pushout topology (second characterization) is determined by specifying that maps from the pushout to any space W are continuous if and only if the maps $X \to W$ and $Y \to W$ obtained by precomposing with f and g are continuous. Said the other way around: maps from the pushout to a space W are specified by maps $a : X \to W$ and $b : Y \to W$ with $af = bg$.

Pushout diagrams like this are commonly used to describe the space obtained by attaching a disc D^n to a space X along a map $f : S^{n-1} \to X$.

$$\begin{array}{ccc} S^{n-1} & \xrightarrow{f} & X \\ {\scriptstyle i}\downarrow & \ulcorner & \downarrow{\scriptstyle i_X} \\ D^n & \xrightarrow{i_Y} & X \sqcup_{S^{n-1}} D^n \end{array}$$

In this case, the map $S^{n-1} \hookrightarrow D^n$ is usually understood as the inclusion, and one describes the pushout succinctly by saying "the disc D^n is attached to X via the attaching map f" and writes $X \sqcup_f D^n$. Some authors (Brown, 2006) call these pushouts "adjunction spaces," but we will not. We reserve "adjunction" for another purpose (see chapter 5).

To summarize, pullbacks provide one way to obtain a limit from the product while pushouts provide one way to obtain a quotient of the coproduct. If we change the shape of the indexing categories, then the (co)limit of the resulting diagrams provide additional constructions: inverse and directed limits.

4.3.4 Inverse and Direct Limits

The limit of a diagram of the shape $\bullet \leftarrow \bullet \leftarrow \bullet \leftarrow \cdots$, such as

$$X_1 \overset{f_1}{\leftarrow} X_2 \overset{f_2}{\leftarrow} X_3 \leftarrow \cdots$$

is sometimes called the *inverse limit* of the objects $\{X_i\}$. As in the case of pullbacks, the inverse limit in Set is a certain subset of the product of the objects. Explicitly, the inverse limit is realized by the set of sequences $(x_1, x_2, \ldots) \in \prod_i X_i$ satisfying $f_i x_{i+1} = x_i$ for all i, together with projection maps from the product down to each factor. It is denoted by $\varprojlim X_i$ and can be thought of as the smallest object that projects down to the factors. The inverse limit in Top has this set of sequences as its underlying set. It becomes a topological space when endowed with the subspace topology of the product. For example, the limit of the diagram of spaces

$$\mathbb{R} \leftarrow \mathbb{R}^2 \leftarrow \mathbb{R}^3 \leftarrow \cdots$$

where the maps $\mathbb{R}^{n+1} \to \mathbb{R}^n$ are given by $(x_1, \ldots, x_n, x_{n+1}) \mapsto (x_1, \ldots, x_n)$ is the product $X = \prod_{n \in \mathbb{N}} \mathbb{R}$, the set of all sequences $(x_1, x_2, x_3, \ldots)$ with the product topology. The projections $X \to \mathbb{R}^n$ defined by $(x_1, x_2, \ldots) \mapsto (x_1, \ldots, x_n)$ define the map from X to the diagram, and the topology on X is the coarsest topology making the maps from X to the diagram continuous. Here, the limit X of the diagram $\mathbb{R} \leftarrow \mathbb{R}^2 \leftarrow \mathbb{R}^3 \cdots$ agrees in both Top and $\mathsf{Vect}_{\mathbf{k}}$.

Dually, the colimit of a diagram of the shape $\bullet \to \bullet \to \bullet \to \cdots$, such as

$$X_1 \to X_2 \to X_3 \to \cdots$$

is sometimes called the *directed limit* of the $\{X_i\}$. It is denoted by $\varinjlim X_i$ and consists of an object X, together with maps $i_k \colon X_k \to X$ that assemble to be a map from the diagram. In a concrete category in which the objects are sets with some additional structure and the $X_k \to X_{k+1}$ are injections, one can think of the diagram as an increasing sequence of objects. The colimit, if it exists, may be thought of as the union of the objects.

As a closing remark, notice that the limit of the diagram $X_1 \to X_2 \to X_3 \to \cdots$ or the colimit of the diagram $X_1 \leftarrow X_2 \leftarrow X_3 \leftarrow \cdots$ are both just the object X_1.

Example 4.1 In linear algebra, the colimit of $\mathbb{N}$ copies of $\mathbb{R}$ is the set of sequences of real numbers for which all but finitely many are zero and is denoted $\oplus_{n \in \mathbb{N}} \mathbb{R}$. This is not the same as the colimit of $\mathbb{N}$ copies of $\mathbb{R}$ in Top, which is $\coprod_{n \in \mathbb{N}} \mathbb{R}$. To make the vector space $\oplus_{n \in \mathbb{N}} \mathbb{R}$ into a topological space, we need to view it in a different way.

Specifically, $X = \oplus_{n \in \mathbb{N}} \mathbb{R}$ is the colimit of the diagram of (vector and topological) spaces

$$\mathbb{R} \to \mathbb{R}^2 \to \mathbb{R}^3 \to \cdots$$

where the map $\mathbb{R}^n \to \mathbb{R}^{n+1}$ is given by $(x_1, \ldots, x_n) \mapsto (x_1, \ldots, x_n, 0)$. Think of the diagram as an increasing union: $\mathbb{R}$ sits inside $\mathbb{R}^2$ as the x axis, then $\mathbb{R}^2$ sits inside $\mathbb{R}^3$ as the xy-plane, and so on. The colimit X of the diagram is an infinite dimensional space in which all these finite dimensional spaces sit inside and is the *smallest* such space, meaning that if Y is any other space that has maps $X_i \to Y$, these maps factor through a map $X \to Y$. The space X is realized as the set of sequences $(x_1, x_2, \ldots)$ for which all but finitely many x_i are nonzero, together with the maps $\mathbb{R}^n \to X$ defined by

$$(x_1, \ldots, x_n) \mapsto (x_1, \ldots, x_n, 0, 0, \ldots)$$

which identify the $\mathbb{R}^n$s with increasing subsets of X. The vector space structure is addition and scalar multiplication of sequences. The topology on X coming from the colimit can be described explicitly by saying that a set U of sequences is open if and only if the intersection $U \cap \mathbb{R}^n$ is open for all $n \in \mathbb{N}$. This is the finest topology that makes the inclusions $R^n \hookrightarrow X$ continuous.

4.3.5 Equalizers and Coequalizers

The limit of a diagram of the shape $\bullet \rightrightarrows \bullet$, such as

$$X \overset{f}{\underset{g}{\rightrightarrows}} Y$$

is called the *equalizer* of f and g. In Set the equalizer is realized as the set $E = \{x \in X \mid fx = gx\}$, together with the inclusion map $E \to X$. It's the largest subset of the domain X on which the two maps agree. In Top, the equalizer has E as its underlying set and becomes a space when endowed with the subspace topology. The universal property is captured in this diagram:

$$S \dashrightarrow E \longrightarrow X \overset{f}{\underset{g}{\rightrightarrows}} Y$$

In algebraic categories, such as Grp, $\mathsf{Vect}_{\mathbb{k}}$, RMod, the equalizer of $f\colon G \to H$ and the unique map from the initial object $0\colon G \to H$ is called the *kernel* of f.

Dually, the colimit of the same diagram

$$X \overset{f}{\underset{g}{\rightrightarrows}} Y$$

is called the *coequalizer* of f and g. In Set and Top, the coequalizer is realized as the quotient $Y/{\sim}$ where $\sim$ is the equivalence relation generated by $fx \sim gx$ for each $x \in X$, endowed with the quotient topology in the case of Top. It's the quotient of the codomain Y by the smallest relation that makes the maps agree. The universal property is captured in this diagram:

$$X \overset{f}{\underset{g}{\rightrightarrows}} Y \longrightarrow C \dashrightarrow S$$

In algebraic categories, such as Grp, $\mathsf{Vect}_{\mathbb{k}}$, RMod, the coequalizer of $f\colon G \to H$ and the unique map from the initial object $0\colon G \to H$ is the called the *cokernel* of f.

After reading through the examples in this chapter, you might suspect that a limit is always, in some sense, either a product or a construction obtained from a product. This suspicion is indeed correct and provides a prescription for constructing limits in general. In fact, it is a theorem: if a category has all products and all equalizers, then it has all limits. Likewise, the feeling that a colimit may be regarded as either a coproduct or a quotient of a coproduct is also a theorem: if a category has all coproducts and all coequalizers, then it has all colimits. We'll close with these results in the next section.

4.4 Completeness and Cocompleteness

A category is called *complete* if it contains the limits of small diagrams and is called *cocomplete* if it contains the colimits of all small diagrams.[1] The categories Set and Top are both complete and cocomplete. In Set, one can construct the colimit of any diagram by taking the disjoint union of every set in the diagram and then quotienting by the relations required for the diagram to map into the resulting set. In Top, this set gets the quotient topology of the disjoint union. This topology is the finest topology for which all the maps involved in the map from the diagram are continuous.

Dually, to construct the limit of any diagram of sets, first take the product of all the sets that appear in the diagram. The product then maps to all objects in the diagram. The limit of the diagram is simply the subset of the product so that the projection maps to the objects assemble to be a map to the diagram. In Top, this set gets the subspace topology of the product. This topology is the coarsest topology for which all the maps involved in the map to the diagram are continuous.

Seeing how to define an arbitrary colimit of sets as a quotient of the disjoint union or how to define an arbitrary limit as a subspace of the product gives the idea of how to prove the following theorem: small (co)limits are all "generated by" a set's worth of (co)products and (co)equalizers.

Theorem 4.1 If a category has products and equalizers, then it is complete. If it has coproducts and coequalizers, then it is cocomplete.

Proof. Here's how to construct the colimit of a diagram in a category with coproducts and coequalizers. Proceed in two steps. First, take the coproduct Y of all the objects X_α in the diagram that have morphisms $X_\alpha \to X_\beta$ from them (there may be multiple copies of X_αs) and take the coproduct Z of all the objects X_β that appear in the diagram (just one copy each):

$$Y := \coprod_{X_\alpha \to X_\beta} X_\alpha \qquad \Bigg| \qquad Z := \coprod_\beta X_\beta$$

There are two maps $Y \to Z$. One map, call it f, is defined by taking the coproduct of the morphisms in the diagram, and one map is defined simply by identities. The coequalizer of these two maps

$$Y \underset{\text{id}}{\overset{f}{\rightrightarrows}} Z$$

[1] A category is *small* if both the collection of objects and the collection of morphisms are sets.

is the colimit of the diagram. The idea for limits is similar. □

A summary of the ideas discussed in this chapter can be organized as seen in table 4.1. We end this chapter with a definition and an example.

Table 4.1 Common categorical limits and colimits.

(index)	$\xrightarrow{\text{functor}}$	(diagram)	its limit	its colimit
	$\longmapsto$		terminal object	initial object
$\bullet \quad \bullet \quad \bullet$	$\longmapsto$	$A \quad B \quad C$	product	coproduct
$\begin{array}{ccc} & & \bullet \\ & & \downarrow \\ \bullet & \longrightarrow & \bullet \end{array}$	$\longmapsto$	$\begin{array}{ccc} & & B \\ & & \downarrow \\ A & \longrightarrow & C \end{array}$	pullback	—
$\begin{array}{ccc} \bullet & \longrightarrow & \bullet \\ \downarrow & & \\ \bullet & & \end{array}$	$\longmapsto$	$\begin{array}{ccc} C & \longrightarrow & B \\ \downarrow & & \\ A & & \end{array}$	—	pushout
$\bullet \leftarrow \bullet \leftarrow \cdots$	$\longmapsto$	$A_1 \leftarrow A_2 \leftarrow \cdots$	inverse limit	—
$\bullet \rightarrow \bullet \rightarrow \cdots$	$\longmapsto$	$A_1 \rightarrow A_2 \rightarrow \cdots$	—	direct limit
$\bullet \rightrightarrows \bullet$	$\longmapsto$	$A \rightrightarrows B$	equalizer	coequalizer

Definition 4.5 A functor is *continuous* if and only if it takes limits to limits. It is *cocontinuous* if it takes colimits to colimits.

Example 4.2 Given a object X in a category C, the hom functor $\mathsf{C}(X, -)\colon \mathsf{C} \to \mathsf{Set}$ is continuous. One may wish for a dual statement, "$\mathsf{C}(-, X)\colon \mathsf{C}^{op} \to \mathsf{Set}$ is cocontinuous," but this is not the case. The contravariance of $\mathsf{C}(-, X)$ does imply that colimits are sent to limits. In fact, we've seen a special case of these results in the discussion of the universal property of products and coproducts in Set in section 0.3.4. An example of a cocontinuous functor may be found in exercise 4.5 at the end of the chapter.

Exercises

1. Let $f\colon Y \to X$ be an embedding of a space Y into a space X. Construct a diagram for which Y (and the map f) is a limit. Hint: exercise 1.12 at the end of chapter 1 shows that quotients are coequalizers and hence colimits.

2. Define the infinite dimensional sphere S^∞ to be the colimit of the diagram

$$S^0 \hookrightarrow S^1 \hookrightarrow S^2 \hookrightarrow S^3 \hookrightarrow \cdots$$

Prove that S^∞ is contractible.

3. From a diagram

$$X \underset{g}{\overset{f}{\rightrightarrows}} Y \xrightarrow{h} Z$$

construct a commutative square

$$\begin{array}{ccc} X \coprod X & \xrightarrow{(f,g)} & Y \\ \downarrow & & \downarrow h \\ X & \longrightarrow & Z \end{array}$$

Prove that the first diagram is a coequalizer precisely if the second is a pushout. Now, from a commutative square

$$\begin{array}{ccc} X & \xrightarrow{f} & Y \\ g\downarrow & & \downarrow p \\ X & \xrightarrow[q]{} & Z \end{array}$$

construct a diagram

$$X \underset{g}{\overset{f}{\rightrightarrows}} X \coprod Y \xrightarrow{(p,q)} Z$$

Prove that the first diagram is a pushout if and only if the second is a coequalizer.

Conclude that a category that has pushouts and coproducts has all colimits. Give a similar argument to prove that a category that has pullbacks and products is closed (Mac Lane, 2013, p. 72, exercise 9).

4. In any category, prove that $f\colon X \to Y$ is an epimorphism if and only if the following square is a pushout:

$$\begin{array}{ccc} X & \xrightarrow{f} & Y \\ f\downarrow & & \downarrow \mathrm{id}_Y \\ Y & \xrightarrow{\mathrm{id}_Y} & Y \end{array}$$

(Mac Lane, 2013, p. 72, exercise 4)

5. For any set X, show that the functor $X \times - : \mathsf{Set} \to \mathsf{Set}$ is cocontinuous (Mac Lane, 2013, p. 118, exercise 4).

6. In a poset, what is the limit, if it exists, of any nonempty diagram? What is the colimit, if it exists?

7. Explain the blanks in table 4.1.

8. Using the image below, prove that a functor $F : \mathsf{B} \to \mathsf{C}$ has a colimit if and only if for all objects $Y \in \mathsf{C}$ there is a natural isomorphism:

$$\mathsf{C}(\operatorname{colim} F, Y) \quad \cong \quad \lim \mathsf{C}(F-, Y)$$

The following provides a guide for the proof. Below on the left, observe that by the universal property of the colimit of F, elements of $\mathsf{C}(\operatorname{colim} F, Y)$ correspond to natural transformations from the diagram F to (the constant functor at) Y. Below on the right, oberve that by the universal property of the limit of the functor $\mathsf{C}(F-, Y) \colon \mathsf{B}^{\mathrm{op}} \to \mathsf{Set}$, the set of all natural transformations from (the constant functor at) the terminal, one-point set $*$ to $\mathsf{C}(F-, Y)$ is the limit, $\lim \mathsf{C}(F-, Y)$. Lastly, in the center we claim the indicated maps form a natural isomorphism:

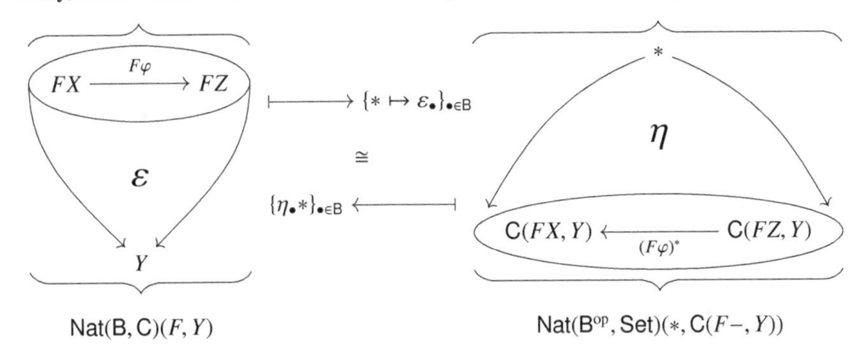

In short, to exchange colimits in the first argument of homs with limits of homs, one need only send maps to precompositions.

5 Adjunctions and the Compact-Open Topology

Birds fly high in the air and survey broad vistas of mathematics out to the far horizon. They delight in concepts that unify our thinking and bring together diverse problems from different parts of the landscape. Frogs live in the mud below and see only the flowers that grow nearby. They delight in the details of particular objects, and they solve problems one at a time.
—Freeman Dyson (2009)

Introduction. Early in chapter 0 it was noted that category theory is an appropriate setting in which to discuss the concept of "sameness" between mathematical objects. This concept is captured by an isomorphism: a morphism from one object to another that is both left and right invertible. The discussion becomes especially interesting when those objects are categories. Two categories are isomorphic if there exists a pair of functors—one in each direction—whose compositions equal the identities. But equality is a lot of ask for! Isomorphisms of categories are too strict to be of much use. Relaxing the situation yields something better: categories C and D are *equivalent* if there exists a pair of functors $L\colon \mathsf{C} \rightleftarrows \mathsf{D}\colon R$ and natural isomorphisms $\mathrm{id}_{\mathsf{C}} \to RL$ and $LR \to \mathrm{id}_{\mathsf{D}}$.

Relaxing this a step further yields another gem of category theory: adjunctions. A pair of functors $L\colon \mathsf{C} \rightleftarrows \mathsf{D}\colon R$ forms an *adjunction*, and L and R are called *adjoint functors*, if there are natural transformations (not necessarily isomorphisms) $\eta\colon \mathrm{id}_{\mathsf{C}} \to RL$ and $\epsilon\colon LR \to \mathrm{id}_{\mathsf{D}}$ that, in addition, interact compatibly in a sense that can be made precise. Here the categories may not be equivalent, but don't think that adjunctions are mere second (or third) best. Quite often, relaxing a notion of equivalence results in a trove of rich mathematics. That is indeed the case here.

In this chapter, then, we introduce adjoint functors and use them to highlight several constructions in topology. We'll present the formal definition in section 5.1 and give some examples—free constructions in algebra, a forgetful functor from Top, and the Stone-Čech compactification—in sections 5.2, 5.3, and 5.5, respectively. Then we'll use a particularly nice adjunction—the product-hom adjunction—as motivation for putting an appropriate topology on function spaces. In section 5.6, we'll take an in-depth look at this topology, called the compact-open topology. Quite a few pages are devoted to this endeavor and some of the difficulties involved. Finally, section 5.7 closes with a discussion on the category of compactly generated weakly Hausdorff spaces—a "convenient" category of topological spaces. So in the pages to come, we'll be both birds and frogs. A categorical point of view oftentimes highlights and elevates the important properties that characterize an object or construction but fails to establish that such objects exist. Existence can require getting down in the mud.

5.1 Adjunctions

As mentioned above, an adjunction consists of a pair of functors L and R and a pair of natural transformations η and ϵ that interact in a certain way. But there is an equivalent definition, which is simpler to digest in a first introduction. We'll present this definition now and jump right into an example. The alternate definition will be given shortly after.

Definition 5.1 Let C and D be categories. An *adjunction* between C and D is a pair of functors $L\colon \mathsf{C} \to \mathsf{D}$ and $R\colon \mathsf{D} \to \mathsf{C}$ together with an isomorphism

$$\mathsf{D}(LX, Y) \overset{\cong}{\longleftrightarrow} \mathsf{C}(X, RY) \tag{5.1}$$

for each object X in C and each object Y in D that is natural in both components. The functor L is called the *left adjoint* and the functor R is called the *right adjoint*. We say the adjunction isomorphism—applied in either direction—sends a morphism to its *adjunct* (or *transpose*) and we write $\hat{f}$ for the adjunct of a map f. Together, all of this information is often denoted by

$$L\colon \mathsf{C} \rightleftarrows \mathsf{D}\colon R$$

or even more succinctly by $L \dashv R$.

To say that the isomorphism in (5.1) is "natural" in both components means that it arises via natural transformations. More precisely, notice that for each object X in C we get a pair of hom-functors

$$\mathsf{D}(LX, -) \qquad\qquad \mathsf{C}(X, R-)$$

from $\mathsf{D} \to \mathsf{Set}$. Similarly, for each object Y in D we have the hom-functors

$$\mathsf{D}(L-, Y) \qquad\qquad \mathsf{C}(-, RY)$$

from $\mathsf{C}^{\mathrm{op}} \to \mathsf{Set}$. Saying the isomorphism $\mathsf{C}(LX, Y) \xrightarrow{\cong} \mathsf{D}(X, RY)$ is "natural in both components" means that there are natural transformations of functors

$$\mathsf{D}(LX, -) \xrightarrow{\quad\cong\quad} \mathsf{C}(X, R-)$$

$$\mathsf{D}(L-, Y) \xrightarrow{\quad\cong\quad} \mathsf{C}(-, RY)$$

Example 5.1 For any sets X, Y, and Z, the bijection $Y^{X\times Z} \xrightarrow{\cong} \left(Y^X\right)^Z$ arises from an adjunction. The functor $X\times -\colon \mathsf{Set} \to \mathsf{Set}$ is left adjoint to the functor $\mathsf{Set}(X, -)\colon \mathsf{Set} \to \mathsf{Set}$. To see it clearly, fix a set X, and define two functors

$$L := X \times -\colon \mathsf{Set} \to \mathsf{Set} \qquad\qquad R := \mathsf{Set}(X, -)\colon \mathsf{Set} \to \mathsf{Set}$$

Then

$$\mathsf{Set}(LZ, Y) = Y^{X\times Z} \cong \left(Y^X\right)^Z = \mathsf{Set}(Z, RY)$$

The setup $L\colon \mathsf{Set} \rightleftarrows \mathsf{Set}\colon R$ is called the *product-hom adjunction*. It will make several appearances in the pages to come. The "hom" in "product-hom" refers to "homomorphisms," which is how people used to, and sometimes still do, refer to "morphisms" in categories. We didn't want to try out a new name, like "product-mor," for this adjunction, however.

5.1.1 The Unit and Counit of an Adjunction

Suppose $L\colon \mathsf{C} \rightleftarrows \mathsf{D}\colon R$ is an adjunction with adjunction isomorphism

$$\varphi_{X,Y}\colon \mathsf{D}(LX, Y) \xrightarrow{\cong} \mathsf{C}(X, RY) \tag{5.2}$$

By setting $Y = LX$, we have an isomorphism

$$\varphi_{X,LX}\colon \mathsf{D}(LX, LX) \xrightarrow{\cong} \mathsf{C}(X, RLX)$$

Under this isomorphism, the morphism id_{LX} in the category D corresponds to a morphism $\eta_X := \varphi_{X,LX}(\mathrm{id}_{LX})\colon X \to RLX$ in the category C. As one can check, these maps assemble into a natural transformation

$$\eta\colon \mathrm{id}_{\mathsf{C}} \to RL$$

called the *unit* of the adjunction. In other words, the unit is comprised of the adjuncts of the identity maps: $\eta_X := \widehat{\mathrm{id}_{LX}}$. Similarly for $X = RY$, under the isomorphism

$$\varphi_{RY,Y}\colon \mathsf{D}(LRY, Y) \xrightarrow{\cong} \mathsf{C}(RY, RY)$$

the morphism id_{RY} in C corresponds to a morphism $\epsilon_Y := \widehat{\mathrm{id}_{RY}}\colon LRY \to Y$ that defines a natural transformation

$$\epsilon\colon LR \to \mathrm{id}_{\mathsf{D}}$$

called the *counit* of the adjunction.

Understanding the counit and unit of an adjunction helps to understand the isomorphisms (5.2) and their corresponding universal properties. For example, suppose $X \in \mathsf{C}$. For any $Y \in \mathsf{D}$ and $f\colon X \to RY$, there exists a unique $g\colon RLX \to RY$ so that $g\eta_X = f$. Here's the picture.

$$\begin{array}{ccc} RLX & & \\ {\scriptstyle \eta_X}\uparrow & \overset{g}{\searrow} & \\ X & \xrightarrow{f} & RY \end{array}$$

Indeed, we can explicitly identify g as $R\hat{f}$. The equality $g\eta_X = f$ follows from the naturality of the adjunction isomorphism $\varphi\colon \mathsf{D}(LX, -) \to \mathsf{C}(X, R-)$.

Explicitly, fix X and Y, and let $f \in \mathsf{C}(X, RY)$. Then we have the adjunct map $\hat{f} \in \mathsf{D}(LX, Y)$ where $\varphi\hat{f} = f$. Moreover, this square commutes:

$$\begin{array}{ccc} \mathsf{D}(LX, LX) & \xrightarrow{\varphi} & \mathsf{C}(X, RLX) \\ \big\downarrow{\scriptstyle \hat{f}_*} & & \big\downarrow{\scriptstyle (R\hat{f})_*} \\ \mathsf{D}(LX, Y) & \xrightarrow{\varphi} & \mathsf{C}(X, RY) \end{array}$$

Choosing $\mathrm{id}_{LX} \in \mathsf{D}(LX, LX)$ and noting that $\varphi\, \mathrm{id}_{LX} = \eta_X$, commutativity implies $\varphi\hat{f} = R\hat{f}\eta$, that is $f = g\eta$ with $g = R\hat{f}$.

Example 5.2 Let's look at the unit and counit of the product-hom adjunction $L\colon \mathsf{Set} \rightleftarrows \mathsf{Set}\colon R$ in Set where

$$L = X \times -\colon \mathsf{Set} \to \mathsf{Set} \qquad \text{and} \qquad R = \mathsf{Set}(X, -)\colon \mathsf{Set} \to \mathsf{Set}$$

The counit of this adjunction is the *evaluation map* $\mathrm{eval}\colon X{\times}Y^X \to Y$ defined by $\mathrm{eval}(x, f) = f(x)$. The unit is the map $Z \to (X \times Z)^X$ defined by $z \mapsto (-, z)$, where $(-, z)\colon X \to X \times Z$ is the function $x \mapsto (x, z)$.

The natural transformations η and ϵ provide another way to define an adjunction.

Definition 5.2 An *adjunction* between categories C and D is a pair of functors $L\colon \mathsf{C} \to \mathsf{D}$ and $R\colon \mathsf{D} \to \mathsf{C}$ together with natural transformations $\eta\colon \mathrm{id}_{\mathsf{C}} \to RL$ and $\epsilon\colon LR \to \mathrm{id}_{\mathsf{D}}$ such that for all objects $X \in \mathsf{C}$ and $Y \in \mathsf{D}$ the following triangles commute:

$$\begin{array}{ccc} LX & \xrightarrow{L\eta_X} & LRLX \\ & {\scriptstyle \mathrm{id}_{LX}} \searrow & \big\downarrow{\scriptstyle \epsilon_{LX}} \\ & & LX \end{array} \qquad\qquad \begin{array}{ccc} RY & \xrightarrow{\eta_{RY}} & RLRY \\ & {\scriptstyle \mathrm{id}_{RY}} \searrow & \big\downarrow{\scriptstyle R\epsilon_Y} \\ & & RY \end{array}$$

Verifying the equivalence of definitions 5.1 and 5.2 is a good exercise.

The next few sections contain additional examples of adjunctions. The first example arises in an algebraic context, while the remaining examples come from topology.

5.2 Free-Forgetful Adjunction in Algebra

Often, free constructions in algebra (free modules, free groups, free abelian groups, free monoids, etc.) are defined by universal properties. To be concrete, let's consider free groups, since modifying the discussion for other free constructions is usually easy. Here's the way a free group is commonly defined:

> *A free group on a set S is a group FS together with a map of sets $\eta\colon S \to FS$ satisfying the property that for any group G and any map of sets $f\colon S \to G$ there exists a unique group homomorphism $\hat{f}\colon FS \to G$ so that $\hat{f}\eta = f$.*

This diagram can help:

$$\begin{array}{ccc} FS & & \\ {\scriptstyle \eta}\uparrow & \overset{\hat{f}}{\searrow} & \\ S & \xrightarrow{f} & G \end{array}$$

and yet, the diagram is also confusing. After all, some objects in the diagrams are sets, some are groups, some of the arrows are set maps, and some are group homomorphisms. The situation becomes clearer when one observes that there is a *forgetful functor* U from groups to sets.

$$\mathsf{Grp} \xrightarrow{U} \mathsf{Set}$$

$$\begin{array}{ccc} G & & UG \\ {\scriptstyle \varphi}\downarrow & \longmapsto & \downarrow{\scriptstyle U\varphi := \varphi} \\ G' & & UG' \end{array}$$

It assigns to any group its underlying set (which explains the letter "U") and to any group homomorphism its underlying function. The adjective *forgetful* is often applied to a functor that "forgets" some or all of the structure of the objects in its codomain. It is a loose term that can be applied to lots of functors whose main job is to drop some data.

One may then define a *free group on a set* S to be a group FS and a map $\eta\colon S \to UFS$ with the property that, for all groups G and maps $f\colon S \to UG$, there exists a unique map $\hat{f}\colon FS \to G$ so that $f = U\hat{f}\eta$. So the right picture is in Set:

$$\begin{array}{ccc} UFS & & \\ {\scriptstyle \eta}\uparrow & \overset{U\hat{f}}{\searrow} & \\ S & \xrightarrow{f} & UG \end{array}$$

Notice that the "there exists" part of the definition of a free group says that for every group G, the map $\mathsf{Grp}(FS, G) \to \mathsf{Set}(S, UG)$ is surjective. The "unique" part of the definition says that $\mathsf{Grp}(FS, G) \to \mathsf{Set}(S, UG)$ is injective. The upshot is that "free" and "forgetful" form an adjoint pair $F\colon \mathsf{Set} \rightleftarrows \mathsf{Grp}\colon U$, providing the isomorphism,

$$\mathsf{Set}(S, UG) \quad \cong \quad \mathsf{Grp}(FS, G)$$

The unit of this adjunction defines the inclusion $\eta\colon S \to UFS$.

Remark 5.1 The universal property defining a free group can also be understood within the context of set maps $S \to UG$ for all groups G. So, one could make a category out of this context. Let's call this category U^S. An object in U^S is a group G and a set map $f\colon S \to UG$. A morphism between two objects $S \xrightarrow{f} UG$ and $S \xrightarrow{f'} UG'$ is a group homomorphism $\varphi\colon G \to G'$ so that $U\varphi f = f'$. That is,

$$\begin{array}{ccc} & S & \\ {}^{f}\swarrow & & \searrow^{f'} \\ UG & \xrightarrow{U\varphi} & UG' \end{array}$$

Then,

a free group on a set S is an initial object in the category U^S.

The context for the universal property is put into a category U^S that is built out of the undisguised material involved: the set S *and* the functor $U\colon \mathsf{Grp} \to \mathsf{Set}$. Then the universal object is a familiar notion (an initial object) from category theory.

The algebraic discussion here arose from considering a forgetful functor. One also has a forgetful functor in the topological setting, which gives rise to further adjunctions.

5.3 The Forgetful Functor U: Top → Set and Its Adjoints

There is a forgetful functor $U\colon \mathsf{Top} \to \mathsf{Set}$ that assigns to any topological space $(X, \mathcal{T}_X)$ the set X and to any continuous function $f\colon (X, \mathcal{T}_X) \to (Y, \mathcal{T}_Y)$ the function $f\colon X \to Y$. It is both a left and right adjoint in Top. Define a functor $D\colon \mathsf{Set} \to \mathsf{Top}$ that assigns to any set X the space $(X, \mathcal{T}_{\text{discrete}})$ with the discrete topology. To any function $f\colon X \to Y$, let $Df = f$, which is a continuous function. The setup $D\colon \mathsf{Set} \rightleftarrows \mathsf{Top}\colon U$ is an adjunction; for any set X and any space Y, we have

$$\mathsf{Top}(DX, Y) \quad \cong \quad \mathsf{Set}(X, UY)$$

On the right, we have arbitrary functions from the set X into the space Y, viewed as a set. On the left, we take continuous functions $DX \to Y$, which are *all* functions from $X \to Y$, since every function from a discrete space is continuous.

But here's another functor $I\colon \mathsf{Set} \to \mathsf{Top}$ that assigns to a set the same set with the indiscrete topology and to any function $f\colon X \to Y$, the same function, which will be continuous. Then $U\colon \mathsf{Top} \rightleftarrows \mathsf{Set}\colon I$ is an adjunction; for any space X and any set Y, we have

$$\mathsf{Set}(UX, Y) \quad \cong \quad \mathsf{Top}(X, IY)$$

On the left we have arbitrary functions from X, viewed as a set, to the set Y. On the right, we have continuous functions $X \to IY$, which are all functions $X \to Y$ since every function into an indiscrete space is continuous.

The universal properties arising from these adjunctions don't seem very interesting, but the fact that U is both a left and a right adjoint has important consequences. Notice this theorem in particular.

Theorem 5.1 If $L\colon \mathsf{C} \to \mathsf{D}$ has a right adjoint, then L is cocontinuous. If $R\colon \mathsf{D} \to \mathsf{C}$ has a left adjoint, then R is continuous.

Proof. Recall from exercise 4.8 at the end of chapter 4 that there is a natural isomorphism:

$$\mathsf{C}(\operatorname{colim} F, Y) \cong \lim \mathsf{C}(F(-), Y)$$

for any functor $F\colon \mathsf{B} \to \mathsf{C}$. Therefore,

$$\begin{aligned} \mathsf{D}(L(\operatorname{colim} F), Y) &\cong \mathsf{C}(\operatorname{colim} F, RY) \\ &\cong \lim \mathsf{C}(F-, RY) \\ &\cong \lim \mathsf{D}(LF-, Y) \\ &\cong \mathsf{D}(\operatorname{colim} LF, Y) \end{aligned}$$

Therefore, $L(\operatorname{colim} F)$ satisfies the universal property of $\operatorname{colim} LF$. And in particular, because colimits (if they exist) are unique up to unique isomorphism,

$$L(\operatorname{colim} F) \cong \operatorname{colim} LF$$

Thus, L is cocontinuous. By a similar argument—which we encourage the reader to verify—right adjoint functors are continuous. □

Corollary 5.1.1 Right adjoints preserve products.

Proof. Immediate. Products are limits. □

This explains why the constructions of products and coproducts, subspaces and quotients, equalizers and coequalizers, and pullbacks and pushforwards in Top must have, as an underlying set, the corresponding construction in Set: if the construction exists in Top then the forgetful functor $U\colon \mathsf{Top} \to \mathsf{Set}$ preserves it!

5.4 Adjoint Functor Theorems

What about a converse to theorem 5.1? Let $R : \mathsf{D} \to \mathsf{C}$ be any functor. Under what conditions will R have a left adjoint? Clearly, R must be continuous. Is the continuity of R sufficient? Not quite. Here's a nice way to think about it. For each object $X \in C$, look at the category R^X whose objects consist of an object Y in D together with a morphism $X \to RY$. A morphism between $f\colon X \to RY$ and $f'\colon X \to RY'$ is a morphism $g\colon Y \to Y'$ such that $(Rg)f = f'$. As in remark 5.1, given a functor $L\colon \mathsf{C} \to \mathsf{D}$, an object $LX \in \mathsf{D}$ satisfying $\mathsf{D}(LX, Y) \cong \mathsf{C}(X, RY)$ for all Y is an initial object in R^X. If there is an initial object LX in R^X for every object X, then they assemble functorially into a left adjoint $L\colon \mathsf{C} \to \mathsf{D}$. In any category, an initial object is the limit of the identity functor. So, R will have a left adjoint if and only if the identity functor on the category R^X has a limit for all objects X. If the category D is complete, then the functor R being continuous implies that R^X is complete. However, even if R^X is complete, the identity functor on R^X is usually not a *small* diagram. So, there are a suite of theorems known as *adjoint functor theorems* that assume the functor R is continuous and the category D is complete and that add some kind of hypothesis allowing one to use the fact that R^X has limits of all small diagrams to prove that the identity functor on R^X has a limit. We don't use any adjoint functor theorems in this book, but it's good to know they exist. Let's state one precisely with the *Solution Set Condition* (Mac Lane, 2013; Freyd, 1969) hypothesis.

The Solution Set Condition A functor $R \colon \mathsf{D} \to \mathsf{C}$ satisfies the *Solution Set Condition* if and only if for every object X in C, there exists a set of objects $\{Y_i\}$ in D and a set of morphisms

$$S = \{f_i \colon X \to RY_i\}$$

so that any $f \colon X \to RY$ factors through some $f_i \in S$ along a morphism $Y_i \to Y$ in D.

The Adjoint Functor Theorem Suppose D is complete and that $R \colon \mathsf{D} \to \mathsf{C}$ is a continuous functor satisfying the Solution Set Condition. Then R has a left adjoint $L \colon \mathsf{C} \to \mathsf{D}$.

For details beyond what we've already said, see the classic reference by Mac Lane (2013) or section 4.6 of the excellent book by Riehl (2016), and for an enlightening treatment of adjunctions including applications, see Spivak (2014). Before moving on, we should say that there are adjoint functor theorems for the existence of right adjoints as well. They suppose that $L \colon \mathsf{C} \to \mathsf{D}$ is a cocontinuous functor from a cocomplete category C, with some "co" version of the Solution Set Condition.

Our discussion of adjoint functor theorems arose from an observation about the forgetful functor on Top and its adjoints. Another notable adjunction in topology arises in a discussion on compactifications.

5.5 Compactifications

Definition 5.3 A *compactification* of a topological space is an embedding of the space as a dense subspace of a compact Hausdorff space.

So a compactification of X is a compact Hausdorff space Y and a continuous injection $i \colon X \to Y$ with $X \cong iX \subseteq Y$ and $\overline{X} = Y$. Note that only Hausdorff spaces have compactifications since every subspace of a Hausdorff space is Hausdorff.

Example 5.3 The inclusion $(0, 1) \hookrightarrow [0, 1]$ and the map $(0, 1) \hookrightarrow S^1$ defined by $t \mapsto (\cos 2\pi t, \sin 2\pi t)$ are both compactifications. For the space $X = (0, 1)$ with the discrete topology, the map $X \hookrightarrow [0, 1]$ is not an embedding and is hence not a compactification.

5.5.1 The One-Point Compactification

If a compactification Y of a space X is obtained by adding a single point to X, then $X \hookrightarrow Y$ is called a *one-point compactification*—also sometimes called the *Alexandroff one-point compactification*. A space X has one-point compactification if and only if X is Hausdorff and locally compact. If a space has a one-point compactification, then it's unique.

To see this, suppose $X \hookrightarrow X^*$ is a compactification and $X^* \setminus X = \{p\}$. The open neighborhoods of p are precisely the complements of compact subsets of X: the complement of an open set containing p is a closed subset of a compact space and so is compact. Conversely if K is a compact subset of $X \subset X^*$, then it is closed, so its complement in X^* is an open set containing p. Then, because points of X can be separated by open sets from the point $p \in X^* \setminus X$, there's a neighborhood of every point of X contained in a compact set, so X is locally compact. The fact that X is a dense subset of X^* implies $\{p\}$ is not open, meaning that X is not already compact.

Conversely, beginning with any space X, one constructs a new space by adding a point p and defining the open neighborhoods of p to be complements of compact sets in X. If X is locally compact and Hausdorff and not compact, the result is a topology on $X^* := X \cup \{p\}$ that is compact and Hausdorff having X as a dense subset.

Now, what kind of property does the one-point compactification have?

Theorem 5.2 Suppose X is locally compact, Hausdorff, and not compact, and let $i\colon X \to X^*$ be the one-point compactification of X. If $e\colon X \to Y$ is any other compactification of X, then there exists a unique quotient map $q\colon Y \to X^*$ with $qe = i$.

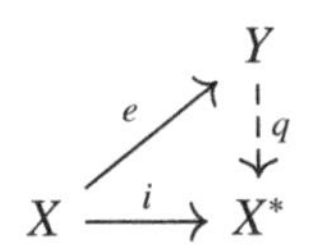

Proof. The idea is that the quotient of Y obtained by identifying $Y \setminus eX$ to one point is homeomorphic to X^*. The details are left as an exercise. □

This theorem can be useful, but it really doesn't say much more than "the one-point compactification of X is the smallest compactification of X," which you may find unsurprising. At the other extreme is the Stone-Čech compactification, which has good categorical properties.

5.5.2 The Stone-Čech Compactification

Let CH be the category whose objects are compact Hausdorff spaces and whose morphisms are continuous functions. There is a functor $U\colon \mathsf{CH} \to \mathsf{Top}$, which is just the inclusion of compact Hausdorff spaces as a subcategory of topological spaces and is the identity on objects and morphisms. The functor U has a left adjoint $\beta\colon \mathsf{Top} \to \mathsf{CH}$ called the *Stone-Čech compactification*. Constructions of β are outlined as construction 6.11 in May (2000) and in more detail in section 38 of Munkres (2000).

For now, let's just unwind this functorial description and see what it means. To say that β is a left adjoint of U means that for every topological space X and every compact Hausdorff space Y, we have a natural bijection

$$\mathsf{CH}(\beta X, Y) \cong \mathsf{Top}(X, UY) = \mathsf{Top}(X, Y)$$

This says continuous functions $f\colon X \to Y$ from a space X to a compact Hausdorff space Y correspond precisely to continuous functions $\hat{f}\colon \beta X \to Y$. Specifying the continuous functions from βX determines the space βX if it exists, but it doesn't *prove* it exists. For that, you need a construction as outlined in the references above, or you could invoke some version of the Adjoint Functor Theorem to prove that $U\colon \mathsf{CH} \to \mathsf{Top}$ has a left adjoint (Mac Lane, 2013).

The unit of the Stone-Čech compactification adjunction

$$\beta\colon \mathsf{Top} \rightleftarrows \mathsf{CH}\colon U$$

defines a morphism $\eta\colon X \to U\beta X$. Since $U\beta X = \beta X$, the Stone-Čech compactification as a left adjoint of U doesn't just produce a compact Hausdorff space βX from any topological space X; it also produces a continuous function $\eta\colon X \to \beta X$ involved in a universal property. For every map $f\colon X \to Y$ between X and a compact Hausdorff space Y, there is a unique map $U\hat{f} = \hat{f}\colon \beta X \to Y$, the adjunct of f, so that $\hat{f}\eta = f$. Pictorally,

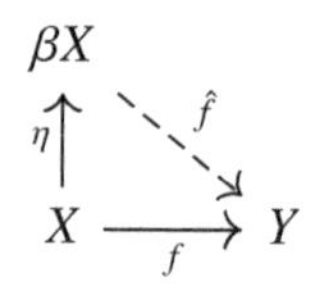

In the case when X is locally compact and Hausdorff, the map $\eta\colon X \to \beta X$ is a compactification of X. That is, $\eta\colon X \to \beta X$ is an embedding and $\overline{X} = \beta X$. Then for any compact Hausdorff space Y, the map $\hat{f}\colon \beta X \to Y$ is the extension of the map $f\colon X \to Y$.

You might have noticed that the triangle above has the same flavor as the triangle defining free groups discussed earlier. This is no coincidence. In addition to the constructions already cited, one can use ultrafilters to construct the Stone-Čech compactification. For any space X, there is a natural topology on the set βX of ultrafilters on a X. The space βX with this topology is compact and Hausdorff, and the inclusion $\eta_X\colon X \to \beta X$ defined by sending a point to its principal ultrafilter is a realization of the Stone-Čech compactification. The fact that there exists an algebraic structure called a *monad* on the ultrafilter functor β sheds further light on resemblance between the Stone-Čech compactification and the free-forgetful adjunctions in algebraic categories mentioned earlier in this chapter. For details, we refer interested readers to the compactum article at the nLab (Stacey et al., 2019) as well as E. Manes's original paper (1969).

In closing, note that unlike the Stone-Čech compactification, the one-point compactification X^* of a locally compact Hausdorff space X is easy to define, but it doesn't have good properties with respect to morphisms. It definitely doesn't satisfy the condition that

$$\mathsf{CH}(X^*, Y) \cong \mathsf{Top}(X, Y)$$

For a simple example, consider $X = (0, 1)$ and its one point compactification $i\colon (0, 1) \to S^1$. Let $Y = [0, 1]$ and consider the inclusion $f\colon (0, 1) \to [0, 1]$. It cannot be extended to a continuous function from $S^1 \to [0, 1]$; there is no diagonal map that fits into the diagram below.

$$\begin{array}{ccc} S^1 & & \\ {\scriptstyle i}\uparrow & & \\ (0,1) & \xrightarrow{f} & [0,1] \end{array}$$

In the next section, we'll continue the discussion of adjunctions in topology. So far, we've discussed free constructions in algebra, the forgetful functor in Top and its adjoints, and compactifications. All of these are are united by the language of adjunctions. Next, we turn to the topic of *mapping spaces*. For any topological spaces X and Y, there is a set of continuous maps $\mathsf{Top}(X, Y)$ between them. Can that set be viewed as a *space* itself? That is, for any $X, Y \in \mathsf{Top}$, can $\mathsf{Top}(X, Y)$ also be regarded as an object in Top in a useful way? We'll see that finding a suitable topology for $\mathsf{Top}(X, Y)$ is more subtle than, say, finding a vector space structure on the set $\mathsf{Vect}_{\mathsf{k}}(V, W)$ of linear maps between vector spaces V and W. In chapter 1, universal properties in Set guided us as we constructed new spaces from old. We will also have categorical guidance on the journey to define topologies on mapping spaces. The guide this time is the product-hom adjunction in Set.

5.6 The Exponential Topology

Let X and Y be spaces. Consider the general problem of equipping the set of continuous functions $\mathsf{Top}(X, Y)$ with a topology making it a space of maps, or a *mapping space*. For the record, the product topology is usually not an appropriate topology for $\mathsf{Top}(X, Y)$ since it treats the space X only as an index set—it doesn't use the topology of X except to identify the continuous functions within the set of all functions $X \to Y$. But what properties should a topology on $\mathsf{Top}(X, Y)$ have? We take as guidance the following desired property.

Desired Property For a fixed space X, the functors

$$X \times -\colon \mathsf{Top} \to \mathsf{Top} \qquad \text{and} \qquad \mathsf{Top}(X, -)\colon \mathsf{Top} \to \mathsf{Top}$$

should form an adjoint pair. That is, for all spaces Y and Z, we should have an isomorphism of sets $\mathsf{Top}(X \times Z, Y) \cong \mathsf{Top}(Z, \mathsf{Top}(X, Y))$.

Let's begin to analyze this property. First think of three fixed spaces X, Y, and Z. One can obtain a function $X \to Y$ by starting with a function $g\colon X \times Z \to Y$ of two variables and by fixing one of the variables $z \in Z$, resulting in $g(-, z)\colon X \to Y$. We'd like a topology on

$\mathsf{Top}(X, Y)$ to have the property that, if the function g of two variables is continuous, then the assignment $z \mapsto g(-, z)$ will define a continuous map $Z \to \mathsf{Top}(X, Y)$. Going the other way, if we have a continuous map $Z \to \mathsf{Top}(X, Y)$, then we should be able to assemble the family of continuous maps from X to Y that are continuously parametrized by the space Z into a single continuous map $X \times Z \to Y$ of two variables.

Now, let's look more closely at the desired property. Let $g\colon X \times Z \to Y$ be continuous. Denote the adjunct by $\hat{g}\colon Z \to \mathsf{Top}(X, Y)$. For $\hat{g}$ to be continuous, the topology on $\mathsf{Top}(X, Y)$ should be rather coarse. However, if the topology on $\mathsf{Top}(X, Y)$ is too coarse (think of the indiscrete topology), then the set $\mathsf{Top}(Z, \mathsf{Top}(X, Y))$ will contain too many continuous functions—it will contain functions that are not the adjunct of any continuous map $g\colon X \times Z \to Y$. It's a balancing act that turns out neatly; if there exists a topology on $\mathsf{Top}(X, Y)$ so that for any space Z, the correspondence $g \mapsto \hat{g}$ defines a bijection of sets

$$\mathsf{Top}(X \times Z, Y) \cong \mathsf{Top}(Z, \mathsf{Top}(X, Y))$$

then that topology is unique (Arens and Dugundji, 1951; Escardó and Heckmann, 2002). Let's call it the *exponential topology* on $\mathsf{Top}(X, Y)$.

This balancing act is reminiscent of the constructions of new spaces from old in chapter 1. That's what we're doing here as well: given two spaces X and Y, we want to make a new topological space from the set of continuous maps from X to Y. In chapter 1, we had guidance from universal properties characterizing similar constructions in Set. Here, the product-hom adjunction in Set provides our categorical guidance. Let's go through some of the details.

Let's call a topology on $\mathsf{Top}(X, Y)$ *splitting* if the continuity of $g\colon Z \times X \to Y$ implies the continuity of $\hat{g}\colon Z \to \mathsf{Top}(X, Y)$. Let's call a topology on $\mathsf{Top}(X, Y)$ *conjoining* if the continuity of $\hat{g}\colon Z \to \mathsf{Top}(X, Y)$ implies the continuity of $g\colon Z \times X \to Y$. Then, to repeat the previous remarks using this terminology (Render, 1993), a topology on $\mathsf{Top}(X, Y)$ must be rather coarse to be splitting and must be rather fine to be conjoining. A topology on $\mathsf{Top}(X, Y)$ is exponential if and only if it is both splitting and conjoining.

Now keep in mind two things. First, the evaluation map is the counit of the product-hom adjunction in Set. Second, the adjunct of the evaluation map is the identity. Together, these give a very nice characterization of conjoining topologies.

Lemma 5.1 A topology on $\mathsf{Top}(X, Y)$ is conjoining if and only if the evaluation map $\mathrm{eval}\colon X \times \mathsf{Top}(X, Y) \to Y$ is continuous.

Proof. Assume we have a topology on $\mathsf{Top}(X, Y)$ for which the evaluation map is continuous. Consider a continuous map $\hat{g}\colon Z \to \mathsf{Top}(X, Y)$, and look at the following diagram

$$X \times Z \xrightarrow{\mathrm{id} \times \hat{g}} X \times \mathsf{Top}(X, Y) \xrightarrow{\mathrm{eval}} Y$$

The identity is continuous, $\hat{g}$ is continuous, and eval is continuous, so the composition eval(id $\times \hat{g}$) is continuous. The composition is precisely g, proving that the topology on $\mathsf{Top}(X,Y)$ is conjoining

For the other direction, assume we have a conjoining topology on $\mathsf{Top}(X,Y)$. Since the adjunct of the evaluation map $\widehat{\text{eval}}\colon \mathsf{Top}(X,Y) \to \mathsf{Top}(X,Y)$ is the identity, which is always continuous, we conclude that the evaluation map is continuous. □

Lemma 5.2 Every splitting topology on $\mathsf{Top}(X,Y)$ is coarser than every conjoining topology.

Proof. Let $\mathcal{T}, \mathcal{T}'$ be topologies on $\mathsf{Top}(X,Y)$. If $\mathcal{T}'$ is conjoining, then the evaluation map $X \times (\mathsf{Top}(X,Y), \mathcal{T}') \to Y$ is continuous. If in addition $\mathcal{T}$ is splitting, then the adjunct of the evaluation map $X \times (\mathsf{Top}(X,Y), \mathcal{T}') \to Y$ is continuous. Since the adjunct of the evaluation map is the identity $(\mathsf{Top}(X,Y), \mathcal{T}') \to (\mathsf{Top}(X,Y), \mathcal{T})$, we conclude that $\mathcal{T} \subseteq \mathcal{T}'$. □

The balancing act follows directly from Lemma 5.2.

Theorem 5.3 If there exists an exponential topology on $\mathsf{Top}(X,Y)$, then it is unique.

Proof. Suppose $\mathcal{T}$ and $\mathcal{T}'$ are exponential topologies on $\mathsf{Top}(X,Y)$. Since $\mathcal{T}$ is splitting and $\mathcal{T}'$ is conjoining, we have $\mathcal{T} \subseteq \mathcal{T}'$. And vice versa: we have $\mathcal{T}' \subseteq \mathcal{T}$ since $\mathcal{T}'$ is splitting and $\mathcal{T}$ is conjoining. □

The catch, as you might have guessed, is that there might not exist an exponential topology on $\mathsf{Top}(X,Y)$—there may be a gap between the splitting topologies and the conjoining topologies on $\mathsf{Top}(X,Y)$. So at this point, you might wonder about taking an "adjoint functor theorem" approach to finding a right adjoint to the functor $X \times -\colon \mathsf{Top} \to \mathsf{Top}$. To do so, we would have to explore the extent to which $X \times -$ preserves colimits. Whether $X \times -$ preserves colimits depends on X. The spaces X for which the functor $X \times -$ preserves colimits were characterized in 1970 as those spaces that are *core-compact* (Day and Kelly, 1970). We'll skip the definition of core-compact here, except to say that a Hausdorff space is core-compact if and only if it is locally compact. It follows that for a locally compact Hausdorff space X, there does exist an exponential topology on $\mathsf{Top}(X,Y)$ for any space Y. Even better, this exponential topology, when X is locally compact Hausdorff, coincides with what is classically called the *compact-open topology* (Fox, 1945). So, at least when X is locally compact and Hausdorff, we can get started thinking about the desired categorical properties of mapping spaces using ideas in classical topology. This is what we do in the next few sections.

Before going on to the compact-open topology, we should give a general categorical definition. In any category C that has finite products, one can ask if for all $X, Y \in \mathsf{C}$ the set $\mathsf{C}(X,Y)$ can be considered as an object in C and if so, whether it supports a product-hom adjunction $X \times -\colon \mathsf{C} \rightleftarrows \mathsf{C}\colon \mathsf{C}(X,-)$. If the answer is yes, then the category is referred to as *Cartesian closed*. The fact that exponential topologies do not always exist implies that the category Top is *not* Cartesian closed. One might try to find a "convenient" category of topological spaces in which the product-hom adjunction holds that is also rich enough to contain the spaces we care about (Brown, 2006; Isbell, 1975; Steenrod, 1967; Stacey et al., 2019). One might guess that locally compact Hausdorff spaces are such a category. But no. Nevermind that we might want some non-Hausdorff spaces. If X and Y are both locally compact and Hausdorff, then $\mathsf{Top}(X, Y)$ may not be. In the final section 5.7 of this chapter, we discuss the search for a convenient Cartesian closed category of topological spaces. That discussion will involve a shift in perspective that, once again, is illuminated with adjunctions.

5.6.1 The Compact-Open Topology

Let's now define the compact-open topology, and try to give you a feel for it.

Definition 5.4 Let X and Y be topological spaces. For each compact set $K \subseteq X$ and each open set $U \subseteq Y$, define $S(K, U) := \{f \in \mathsf{Top}(X, Y) \mid fK \subseteq U\}$. The sets $S(K, U)$ form a subbasis for a topology on $\mathsf{Top}(X, Y)$ called the *compact-open topology*.

Notice that a subbasis for the product topology on $\mathsf{Top}(X, Y)$ consists of sets

$$S(F, U) = \{(f\colon X \to Y) \mid fF \subseteq U\}$$

where $F \subseteq X$ is finite and $U \subseteq Y$ is open. That is, the product topology is what one might call the "finite-open" topology. In the case when X has the discrete topology, all functions $X \to Y$ are continuous and the compact-open topology on $\mathsf{Top}(X, Y)$ coincides with the product topology on $\mathsf{Top}(X, Y)$. More generally, every finite set is compact, so the compact-open topology is finer than the product-topology. As a consequence, fewer filters converge in the compact-open topology than in the product topology.

In fact, a closer look at convergence can give you a good feel for the difference between the product topology and the compact-open topology. Let's start by looking at sequences. A sequence of functions $\{f_n\colon [0, 1] \to [0, 1]\}_{n \in \mathbb{N}}$ converges to a limiting function f in the product topology if and only if the sequence converges pointwise. On the other hand, a sequence of functions $\{f_n\colon [0, 1] \to [0, 1]\}_{n \in \mathbb{N}}$ converges to f in the compact-open topology if and only if the sequence converges uniformly. To see this, consider a more general situation. Suppose that X is compact and Y is a metric space. Then $\mathsf{Top}(X, Y)$ becomes a metric space with the metric defined by

$$d(f, g) := \sup_{x \in X} d(fx, gx)$$

Two functions $f, g \in \mathsf{Top}(X, Y)$ are close in this metric if their values fx and gx are close for all points $x \in X$. A sequence $\{f_n\}$ in $\mathsf{Top}(X, Y)$ converges to f in this metric topology if and only if for all $\varepsilon > 0$ there exists an $n \in N$ so that for all $k > n$ and for all $x \in X$,

$d(f_k x, g_k x) < \varepsilon$. The fact that when X is compact and Y is a metric space the compact-open topology coincides with this metric topology is the content of the next theorem. First, a lemma.

Lemma 5.3 Let X be a metric space and let U be open. For every compact set $K \subseteq U$, there is an $\varepsilon > 0$ so that for any $x \in K$ and any $y \in X \setminus U$, $d(x, y) > \varepsilon$.

Proof. This is a straightforward argument using the definition of compactness. □

Theorem 5.4 Let X be compact and Y be a metric space. The compact-open topology on $\mathsf{Top}(X, Y)$ is the same as the metric topology.

Proof. Let $f \in \mathsf{Top}(X, Y)$ and $\varepsilon > 0$ be given. Consider $B(f, \varepsilon)$. We'll find a set O that is open in the compact-open topology, with $f \in O \subseteq B(f, \varepsilon)$. Hence, compact-open neighborhoods of f refine the metric neighborhoods of f, proving that the compact-open topology is finer than the metric topology. Now, since X is compact, its image fX is compact. Since the collection $\left\{B\left(fx, \frac{\varepsilon}{3}\right)\right\}_{x \in X}$ is an open cover of fX it has a finite subcover

$$\left\{B\left(fx_1, \tfrac{\varepsilon}{3}\right), \ldots, B\left(fx_n, \tfrac{\varepsilon}{3}\right)\right\}$$

Define compact subsets $\{K_1, \ldots, K_n\}$ of X and open subsets $\{U_1, \ldots, U_n\}$ of Y by

$$K_i := \overline{f^{-1}\left(B\left(fx_i, \tfrac{\varepsilon}{3}\right)\right)} \qquad \text{and} \qquad U_i := B\left(fx_i, \tfrac{\varepsilon}{2}\right)$$

Since f is continuous, $f\overline{A} \subseteq \overline{fA}$ for any set A. In particular,

$$fK_i \subseteq \overline{B\left(fx_i, \tfrac{\varepsilon}{3}\right)} \subseteq B\left(fx_i, \tfrac{\varepsilon}{2}\right) = U_i$$

for each $i = 1, \ldots, n$. Therefore, f is in the open set $O := \cap_{i=1}^n S(K_i, U_i)$. To see that $O \subseteq B(f, \varepsilon)$, let $g \in O$. If $x \in K_i$ for some i, we have $fx, gx \in U_i$ since $f, g \in S(K_i, U_i)$. Therefore,

$$d(fx, gx) \leq d(fx, fx_i) + d(fx_i, gx) = \tfrac{\varepsilon}{2} + \tfrac{\varepsilon}{2} = \varepsilon$$

Since the balls $\left\{B\left(fx_i, \frac{\varepsilon}{3}\right)\right\}$ cover fX, the compact sets $\{K_i\}$ cover X and every point x lies in K_i for some i. Therefore, $d(fx, gx) < \varepsilon$ for every $x \in X$, and so $d(f, g) < \varepsilon$ in $\mathsf{Top}(X, Y)$.

To show that the metric topology is finer than the compact-open topology, let $K \subseteq X$ be compact, $U \subseteq Y$ be open, and consider $f \in S(K, U)$. From Lemma 5.3, we know there exists a fixed $\varepsilon > 0$ so that for any $y \in fK$ and any $y' \in Y \setminus fU$, $d(y, y') \geq \varepsilon$. Then if $g \in B(f, \varepsilon)$, we have $d(fx, gx) < \varepsilon$ for every $x \in X$. Therefore, if $x \in K$, then $gx \in U$, and we see that $gK \subseteq U$. This proves $B(f, \varepsilon) \subseteq S(K, U)$. If $O = S(K_1, U_1) \cap \cdots \cap S(K_n, U_n)$ is any basic open set in the compact-open topology, we have the open metric ball $B(f, \varepsilon) \subseteq O$ where $\varepsilon = \min\{\varepsilon_1, \ldots, \varepsilon_n\}$. This proves that every basic open set in the compact-open topology is open in the metric topology and hence the metric-topology is finer than the compact-open topology. □

Now that we've given a feel for the compact-open topology, let's look at it in the context of mapping spaces. The first thing to notice is that the compact-open topology is coarse enough to be splitting.

Theorem 5.5 For any spaces X and Y, the compact-open topology on $\mathsf{Top}(X, Y)$ is splitting.

Proof. Let Z be any space, and suppose $g\colon X \times Z \to Y$ is continuous. To show that the adjunct $\hat{g}\colon Z \to \mathsf{Top}(X, Y)$ is continuous, consider a subbasic open set $S(K, U)$ in $\mathsf{Top}(X, Y)$. We need to show that $(\hat{g})^{-1}S(K, U) = \{z \in Z \mid g(K, z) \subseteq U\}$ is open in Z. Let $z \in (\hat{g})^{-1}S(K, U)$. So, $z \in Z$ and $g(K, z) \subseteq U$. Since g is continuous, we know that $g^{-1}U = \{(x, z) \mid g(x, z) \subseteq U\}$ is open in $X \times Z$ and contains $K \times \{z\}$. Therefore, the Tube Lemma says there are open sets V and W with $K \subseteq V$ and $z \in W$ with $K \times \{z\} \subseteq V \times W \subseteq g^{-1}U$. Then, $z \in W \subseteq (\hat{g})^{-1}S(K, U)$ as needed. □

Theorem 5.6 If X is locally compact and Hausdorff and Y is any space, then the compact-open topology on $\mathsf{Top}(X, Y)$ is exponential.

Proof. We only need to check that the compact-open topology is conjoining, and this is equivalent to showing that the evaluation map $\mathrm{eval}\colon X \times \mathsf{Top}(X, Y) \to Y$ is continuous at every point (x, f). Let $(x, f) \in X \times \mathsf{Top}(X, Y)$, and let $U \subseteq Y$ be an open set containing the evaluation $\mathrm{eval}(x, f) = fx$. Because f is continuous, $f^{-1}U$ is an open set in X containing x. Since X is locally compact and Hausdorff, there exists an open set $V \subseteq X$ with $K := \overline{V}$ compact and $x \in V \subseteq K \subseteq f^{-1}U$. This implies that $fx \in fK \subseteq U$. Then $V \times S(K, U)$ is an open set in $X \times \mathsf{Top}(X, Y)$ with $(x, f) \in V \times S(K, U)$ and $\mathrm{eval}(V \times S(K, U)) \subseteq U$. □

And thus, when X is locally compact and Hausdorff and $\mathsf{Top}(X, Y)$ is equipped with the compact-open topology, we have the desired property listed at the opening of section 5.6. As an application, let's prove theorem 2.20 from chapter 2. We begin with a lemma.

Lemma 5.4 If $f\colon X \to Y$ is a quotient map and Z is locally compact and Hausdorff, then $f \times \mathrm{id}_Z\colon X \times Z \to Y \times Z$ is a quotient map.

Proof. Let $f\colon X \to Y$ be a quotient map. We want to prove that the product $Y \times Z$ has the quotient topology inherited from the map $f \times \mathrm{id}_Z$. So consider $Y \times Z$ with two possibly distinct topologies: $(Y \times Z)_p$ will denote the product topology, and $(Y \times Z)_q$ will denote the quotient topology inherited from the map $f \times \mathrm{id}_Z\colon X \times Z \to Y \times Z$.

The universal property of the quotient topology tells us which maps out of $(Y \times Z)_q$ are continuous. In particular, $\mathrm{id}\colon (Y \times Z)_q \to (Y \times Z)_p$ is continuous since $f \times \mathrm{id}_Z\colon X \times Z \to (Y \times Z)_p$ is continuous. That is, in the diagram below, the dashed map is continuous because the solid diagonal is continuous.

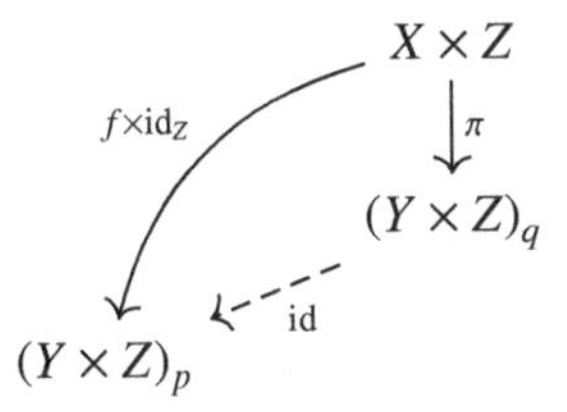

So we only have to prove that the identity in the other direction id: $(Y \times Z)_p \to (Y \times Z)_q$ is continuous. Since Z is locally compact Hausdorff, it suffices to prove that the adjunct $\widehat{\mathrm{id}}\colon Y \to \mathsf{Top}(Z, (Y \times Z)_q)$ is continuous. As a map out of Y, $\widehat{\mathrm{id}}$ will be continuous if its precomposition with the quotient map f is continuous. That is, in the diagram below, the dashed map will be continuous if the solid diagonal map is continuous.

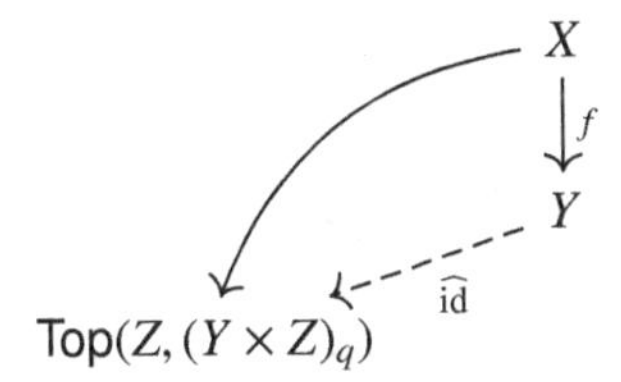

The solid diagonal map in the picture is $\widehat{\pi}$, the adjunct of the continuous quotient map $\pi\colon X \times Z \to (Y \times Z)_q$, and so it is indeed continuous. □

Theorem 5.7 If $X_1 \twoheadrightarrow Y_1$ and $X_2 \twoheadrightarrow Y_2$ are quotient maps and Y_1 and X_2 are locally compact and Hausdorff, then $X_1 \times X_2 \twoheadrightarrow Y_1 \times Y_2$ is a quotient map.

Proof. Suppose Y_1 and X_2 are locally compact and Hausdorff and that $f_1\colon X_1 \twoheadrightarrow Y_1$ and $f_2\colon X_2 \twoheadrightarrow Y_2$ are quotient maps. By the lemma, the two maps $f_1 \times \mathrm{id}_{X_2}\colon X_1 \times X_2 \twoheadrightarrow Y_1 \times X_2$ and $\mathrm{id}_{Y_1} \times f_2\colon Y_1 \times X_2 \twoheadrightarrow Y_1 \times Y_2$ are quotient maps. Therefore the composition

$$(\mathrm{id}_{Y_1} \times f_2) \circ (f_1 \times \mathrm{id}_{X_2}) : X_1 \times X_2 \twoheadrightarrow Y_1 \times Y_2$$

is a quotient map. □

In the next section, we'll give an application to analysis. Unlike $\mathbb{R}^n$, whose compact subsets are nicely characterized by the *Heine-Borel theorem* as the sets that are closed and bounded, it's not always easy to decide when a subset of $\mathsf{Top}(X, Y)$ is compact. Ascoli's theorem in the next section provides a criterion.

5.6.2 The Theorems of Ascoli and Arzela

It's not difficult to decide when a family of functions is compact using the product topology.

Theorem 5.8 If X is any space and Y is Hausdorff, then a subset $A \subseteq \mathsf{Top}(X, Y)$ has compact closure in the product topology if and only if for each $x \in X$, the set $A_x = \{fx \in Y \mid f \in A\}$ has compact closure in Y.

Proof. This was exercise 2.19 at the end of chapter 2. □

If we can identify families of functions for which the product topology and the compact-open topology coincide, then we have necessary and sufficient conditions for such families to be compact in the compact-open topology. The following definition provides a common way to make such an identification.

Definition 5.5 Let X be a topological space and (Y, d) be a metric space. A family $A \subseteq \mathsf{Top}(X, Y)$ is called *equicontinuous at* $x \in X$ if and only if for every $\varepsilon > 0$, there exists an open neighborhood U of x so that for every $u \in U$ and for every $f \in A$, $d(fx, fu) < \varepsilon$. If $\mathcal{F}$ is equicontinuous for every $x \in X$, the family A is simply called *equicontinuous*.

So a family of functions is equicontinuous if, within a neighborhood, one can bound the variation of every function in the family by a single epsilon. For this section, the most important facts about equicontinuous families are that the compact-open topology agrees with the product topology on them and that their closures are also equicontinuous. We leave the proofs of these two facts as exercises for you to solve.

Lemma 5.5 Let X be a topological space and (Y, d) be a metric space. If $A \subseteq \mathsf{Top}(X, Y)$ is an equicontinuous family, then the subspace topology on A of $\mathsf{Top}(X, Y)$ with compact-open topology is the same as the subspace topology on A of $\mathsf{Top}(X, Y)$ with the product topology.

Lemma 5.6 If $A \subseteq \mathsf{Top}(X, Y)$ is equicontinuous, then the closure of A in $\mathsf{Top}(X, Y)$ using the product topology is also equicontinuous.

Putting these ideas together gives us the famous theorems of Ascoli and Arzela.

Ascoli's Theorem Let X be locally compact Hausdorff, and let (Y, d) be a metric space. A family $\mathcal{F} \subseteq \mathsf{Top}(X, Y)$ has compact closure if and only if $\mathcal{F}$ is equicontinuous and for every $x \in X$, the set $F_x := \{fx \mid f \in \mathcal{F}\}$ has compact closure.

Arzela's Theorem Let X be compact, (Y, d) be a metric space and $\{f_n\}$ be a sequence of functions in $\mathsf{Top}(X, Y)$. If $\{f_n\}$ is equicontinuous and if for each $x \in X$ the set $\{f_n x\}$ is bounded, then $\{f_n\}$ has a subsequence that converges uniformly.

5.6.3 Enrich the Product-Hom Adjunction in Top

In the case that $\mathsf{Top}(X, Y)$ has an exponential topology, we know it's unique. Let's introduce some notation for it.

Definition 5.6 Define the space Y^X to be the set $\mathsf{Top}(X, Y)$ with its exponential topology, provided it exists.

Suppose we have an exponential topology on $\mathsf{Top}(X, Y)$ and use Y^X to denote it. We have a bijection of sets

$$\mathsf{Top}(Z \times X, Y) \cong \mathsf{Top}(Z, Y^X) \tag{5.3}$$

Now, we can put the compact-open topology on these mapping spaces. One can then ask the question: under what conditions is the bijection of sets in (5.3) a homeomorphism? One answer is: when X is locally compact and Hausdorff and when Z is Hausdorff. Instead of proving this (as done in Hatcher, 2002, 529–532), let's prove a slightly weaker version by assuming that Z is, additionally, locally compact. Then the compact open topology $\mathsf{Top}(Z, Y^X)$ will be exponential. Also, since the product of locally compact Hausdorff spaces is locally compact Hausdorff (exercise 2.9 at the end of chapter 2), the compact open topology on $\mathsf{Top}(Z \times X, Y)$ will be exponential. So, we're looking at proving that

$$Y^{Z \times X} \cong \left(Y^X\right)^Z$$

The payoff in proving the weaker version is that we have a lot more adjunctions available and we can give a clean, categorical argument, as in Strickland (2009), without diving in to the topologies themselves.

Theorem 5.9 If X and Z are locally compact Hausdorff, then for any space Y, the isomorphism of sets $\mathsf{Top}(Z \times X, Y) \to \mathsf{Top}(Z, \mathsf{Top}(X, Y))$ is a homeomorphsim of spaces.

Proof. Throughout this proof, remember that for any spaces A and C, the compact open topology on $\mathsf{Top}(A, C)$ is splitting. This means that the adjunct of a continuous map $A \times B \to C$, which is a function $B \to \mathsf{Top}(A, C)$, is also continuous. Also, if B is locally compact and Hausdorff, then for any space C, the compact open topology on $\mathsf{Top}(B, C)$ is conjoining. This is equivalent to the statement that for any space C, the evaluation map $B \times \mathsf{Top}(B, C) \to \mathsf{Top}(C)$ is continuous.

Now suppose X and Z are locally compact Hausdorff. Because X is locally compact Hausdorff, the evaluation map $X \times (Y^Z)^X \to Y^Z$ is continuous. Also, because Z is locally compact Hausdorff, the evaluation map $Z \times Y^Z \to Y$ is continuous. Therefore, the composition

$$Z \times X \times (Y^Z)^X \to Z \times Y^Z \to Y$$

is continuous. Now, to keep things clear, set $A = Z \times X$ and set $B = (Y^Z)^X$. We have a continuous function $g \colon A \times B \to Y$. Since the compact-open topology is always splitting, the adjunct $\hat{g} \colon B \to \mathsf{Top}(A, Y)$ of the map g is continuous. Restoring $A = Z \times X$ and

$B = (Y^Z)^X$ reveals that the map $\hat{g}\colon (Y^Z)^X \to \mathsf{Top}(Z \times X, Y)$ is continuous. This map $\hat{g}$ is precisely the map that we want to show is a homeomorphism. We've shown it's continuous, so we're halfway done.

Because $Z \times X$ is locally compact Hausdorff, the evaluation map

$$Z \times X \times Y^{Z \times X} \to Y$$

is continuous. Now, thinking of this as a continuous map from $Z \times -$, its adjunct $X \times Y^{Z \times X} \to Y^Z$ is continuous. And again. Finally, thinking of this as a continuous map from $X \times -$, we conclude its adjunct $Y^{Z \times X} \to (Y^Z)^X$ is continuous. This is the inverse of the continuous map $\hat{g}$ in the previous paragraph, completing the proof. □

Now, let's make some concluding remarks about the compact-open topology on $\mathsf{Top}(X, Y)$ when X is locally compact Hausdorff. From a certain point of view, it's insufficient to work with locally compact Hausdorff spaces. For example, these spaces are not closed under many common constructions. The colimit of the following diagram of locally compact Hausdorff spaces

$$\mathbb{R} \hookrightarrow \mathbb{R}^2 \hookrightarrow \mathbb{R}^3 \hookrightarrow \cdots$$

is not locally compact. Moreover, even if X is locally compact and Hausdorff, Y^X with the compact-open topology may not be locally compact and Hausdorff, so the construction of a topology on a mapping space is not repeatable. One solution to all these issues involves more adjunctions—"k-ification" and "weak-Hausdorfication." This is the next topic.

5.7 Compactly Generated Weakly Hausdorff Spaces

Here, we present a bird's eye view of constructing a topology on $\mathsf{Top}(X, Y)$ in a more general setting. The main idea is to find a "convenient category" of topological spaces which has limits and colimits, has exponential objects with the desirable property listed in section 5.6, and is large enough to contain the spaces we care about. The category of compactly generated weakly Hausdorff spaces is such a category. The importance of compactly generated spaces for topologies on function spaces was recognized early on by Brown (1964). A categorical perspective along the lines we present here, including the behavior of compactly generated spaces under limits, occurred later (Steenrod, 1967). Finally, Hausdorff was replaced with weakly Hausdorff yielding what is often considered the most convenient category of topological spaces (McCord, 1969).

Be aware that the terminology is not used consistently in the literature. For example, "compactly generated" in May (1999) means what we call "compactly generated and weak Hausdorff." Here we err on over-adjectivized terminology in an effort to avoid any confusion, especially since we omit most of the proofs. We'll start with a few definitions.

Definition 5.7 A space X is *compactly generated* if and only if for all compact spaces K and continuous maps $f: K \to X$, the set $f^{-1}A$ being closed (or open) implies A is closed (or open).

Definition 5.8 A space X is *weakly Hausdorff* if and only if for all compact spaces K and continuous maps $f: K \to X$, the image fK is closed in X.

Definition 5.9 Let CG, WH, and CGWH denote the full subcategories of compactly generated, weakly Hausdorff, and compactly generated weakly Hausdorff spaces, respectively.

Example 5.4 The category CG includes all locally compact spaces and all first countable spaces. The category CGWH includes locally compact Hausdorff spaces and metric spaces. Notice that weakly Hausdorff lies between T_1 and T_2. The interested reader will find these properties are all distinct by searching for examples in Steen and Seebach (1995).

The reader might recall from chapter 1 that we introduced constructions of topologies in three ways: first, with a classic definition, which described the open sets; second, with a better definition that characterized the topology among a set of possible topologies; and third, with a still better definition, which characterized which functions into or out of the construction are continuous. Let's describe k-ification in these three ways. Let X be any space. First, one can define a topology on X to consist of all sets U so that $f^{-1}U$ is open for some continuous $f: K \to X$ from a compact space K. This topology is compactly generated. In fact, it is the smallest compactly generated topology containing the original topology on X. It is further characterized by the property that a function $g: X \to Y$ out of X with this topology is continuous if and only for all compact K and all functions $f: K \to X$, $gf: K \to Y$ is continuous. The set X with this compactly generated topology is called the *k-ification* of X and is denoted by kX. The k-ification of a map $f: X \to Y$ is the map f viewed as a map $kX \to kY$, which will be continuous. Thus, k-ification defines a functor $k: \mathsf{Top} \to \mathsf{CG}$.

Theorem 5.10 The following setup is an adjunction: $U: \mathsf{CG} \rightleftarrows \mathsf{Top}: k$, where U is the inclusion of $\mathsf{CG} \to \mathsf{Top}$ and k is the k-ification functor.

Proof. Proved as theorem 3.2 in Steenrod (1967). □

Theorem 5.1 then implies that k preserves limits and U preserves colimits. The statement that k preserves limits implies that the limit of a diagram in Top is sent by k to the limit of the k-ification of the diagram. To clarify what this means, consider two compactly generated spaces X and Y. The product $X \times Y$ in Top may not be compactly generated, but $k(X \times Y)$ is the product of X and Y in CG. This means that $k(X \times Y)$ satisfies the universal property to be a product for compactly generated spaces. The quantifier in the universal property for the product $k(X \times Y)$ involves fewer spaces, and so in general, the topology on $k(X \times Y)$ is finer than the topology on $X \times Y$.

Now, the consequences of U being a left adjoint means that U preserves colimits. This means that *if* a diagram in CG has a colimit in CG, *then* it must agree with the colimit of the diagram in Top. While there's no obvious reason that colimits in CG need exist, it is nevertheless true.

Theorem 5.11 CG is a cocomplete category.

Proof. See appendix A of Lewis (1978). □

Now let's add weakly Hausdorff to the picture with an analogous "weak-Hausdorffification" functor q. Let X be any space. We define the weakly Hausdorff space qX to be the quotient of X by the smallest closed equivalence relation in $X \times X$. You can think about what the open sets of qX are and how to characterize which functions into or out of (which will it be?) qX are continuous.

Theorem 5.12 There is an adjunction $q\colon \mathsf{CG} \rightleftarrows \mathsf{CGWH}\colon U$ where U is the inclusion of $\mathsf{CGWH} \to \mathsf{CG}$.

Proof. See appendix A of Lewis (1978). □

As a consequence, take a diagram in CGWH. The colimit of this diagram may not be weakly Hausdorff, but it is compactly generated by theorem 5.11. Then apply the functor q, which as a left adjoint preserves colimits and hence yields a space in CGWH that must be the colimit of the diagram in CGWH.

As for limits, *if* a diagram in CGWH has a limit in CGWH, then it must agree with the limit of the diagram in CG. While there's no obvious reason that limits in CGWH need exist, it is nevertheless true.

Theorem 5.13 CGWH is a complete category.

Proof. See proposition 2.22 in Strickland (2009). □

The upshot of this back and forth game between adjoint functors produces a category CGWH that is closed under limits and colimits. But be careful of multiple interpretations. For example, for two compactly generated weakly Hausdorff spaces X and Y, we have the "old" product, denoted by $X \times_o Y$, which is the product in Top. We also have the new product, denoted by $X \times Y$, which is the product in CGWH.

Now, for compactly generated weakly Hausdorff spaces X and Y, let $Y^X = k\mathsf{Top}(X, Y)$. That is, the topology we put on the space of maps from X to Y is the k-ification of the compact-open topology.

Theorem 5.14 If X and Y are CGWH, then Y^X is in CGWH. For a fixed X, the assignment $Y \mapsto Y^X$ defines a functor $-^X\colon \mathsf{CGWH} \to \mathsf{CGWH}$ that fits into the adjunction

$$X \times -\colon \mathsf{CGWH} \rightleftarrows \mathsf{CGWH}\colon -^X$$

inducing homeomorphisms of spaces $Y^{X\times Z} \cong (Y^X)^Z$.

Proof. See Lewis (1978). □

Corollary 5.14.1 For $X, Y, Z \in \mathsf{CGWH}$,

(i) the functor $- \times X$ preserves colimits.

(ii) the functor $-^X$ preserves limits.

(iii) the functor Y^- takes colimits to limits.

(iv) composition $Z^Y \times Y^X \to Z^X$ is continuous.

(v) evaluation $\text{eval}\colon X \times Y^X \to Y$ is continuous.

Thus, the category CGWH is Cartesian closed. The category of compactly generated weakly Hausdorff spaces has other good properties, too. We recommend the excellent notes in Strickland (2009) for statements and proofs.

We opened this chapter with the idea that relaxing a notion of equivalence can result in rich mathematics. As we've seen, the data of an adjunction $L \dashv R$ is similar to, but not quite the same as, the data of an equivalence between categories C and D. This relaxed version provides a characterization about relationships between objects in these categories: if you know all morphisms $LX \to Y$ in D, then you know all morphisms $X \to RY$ in C, and vice versa. When the category of interest is Top and X is a locally compact Hausdorff space, the compact-open topology provides a bijection of mapping spaces $\mathsf{Top}(X \times Z, Y) \cong \mathsf{Top}(Z, \mathsf{Top}(X, Y))$ for all spaces Y and Z. More generally, the functors $X \times -$ and $\mathsf{C}(X, -)$ form an adjoint pair when the category C is a convenient one, such as CGWH.

In the next chapter, we'll again use a loosened-up version of equivalence—*homotopy equivalence*—to motivate other rich constructions and adjunctions in topology. As introduced in chapter 1, a homotopy between functions $f, g\colon X \to Y$ is a map $h\colon I \times X \to Y$ with $h(0, -) = f$ and $h(1, -) = g$. Since the unit interval is locally compact and Hausdorff, there is a bijection $\mathsf{Top}(I \times X, Y) \cong \mathsf{Top}(X, \mathsf{Top}(I, Y))$ for all spaces X and Y. On the left are homotopies and on the right are maps into $\mathsf{Top}(I, Y)$, the space of all paths in Y. This dual perspective, together with the ideas in this chapter, naturally leads to a discussion on homotopy, path spaces, and more adjunctions that arise in topology.

Exercises

1. Prove that definitions 5.1 and 5.2 are equivalent.

2. Let $L\colon \mathsf{C} \rightleftarrows \mathsf{D}\colon R$ be an adjunction with unit η and counit ϵ. Let C' be the full subcategory of C whose objects are those $X \in \mathsf{C}$ for which η_X is an isomorphism. Define D' similarly. Show that C' and D' are equivalent categories.

3. Give examples and justify your answers.

a) Find a space X and a space Y for which the evaluation map $\mathsf{Top}(X, Y) \times X \to Y$ is not continuous.

b) Find a space X and a space Y for which the evaluation map $\mathsf{Top}(X, Y) \times X \to Y$ is not continuous.

4. Prove Lemmas 5.5 and 5.6

5. Let $A = \{f \in \mathsf{Top}([0, 1], [0, 1]) \mid f \text{ is differentiable and } |f'| \leq 1\}$. Prove that $\overline{A}$ is compact.

6. Define a family of functions $\mathcal{F} \subset \mathsf{Top}([0, 1], \mathbb{R})$ by $\mathcal{F} = \{f_a\}_{0<a\leq 1}$ where $f_a(x) = 1 - \frac{x}{a}$. Prove or disprove: $\mathcal{F}$ is compact in the compact-open topology.

7. If Y is a subspace of a space X, then a map $r\colon X \to Y$ is called a *retract* of X onto Y, and Y is said to be a retract of X if $ri = \mathrm{id}_Y$ where $i\colon Y \to X$ is the inclusion map. Let $\beta\colon \mathsf{Top} \to \mathsf{CH}$ be the Stone-Čech compactification functor. Prove that any compact Hausdorff space X is a retract of βUX where $U\colon \mathsf{CH} \to \mathsf{Top}$ includes CH as a subcategory of Top.

8. Suppose that Y is locally compact Hausdorff. Prove that for any spaces X and Z, composition $\mathsf{Top}(X, Y) \times \mathsf{Top}(Y, Z) \to \mathsf{Top}(X, Z)$ is continuous.

9. Show that a space X is Hausdorff if and only if the diagonal D is closed in $X \times_o X$. Here $X \times_o X$ means the product topology. Show that a space X is Hausdorff if and only if the diagonal D is closed in $X \times X$, where $X \times X$ means $k(X \times_o X)$.

10. Complete the proof of theorem 5.1. That is, mimic the proof that left adjoints preserve colimits, substituting a right adjoint and limit. This essentially boils down to finding an analogous natural isomorphism, allowing you to "pull a limit in the second argument of a hom out to a limit of homs."

6 Paths, Loops, Cylinders, Suspensions, . . .

In certain situations (such as descent theorems for fundamental groups à la van Kampen) it is much more elegant, even indispensable for understanding something, to work with fundamental groupoids with respect to a suitable packet of base points....
—Alexander Grothendieck (1997)

Introduction. Chapter 0 introduced the categorical theme that *objects are completely determined by their relationships with other objects*. It has origins in theorem 0.1, a corollary of the Yoneda lemma stating that objects X and Y in a category are isomorphic if and only if the corresponding sets $\mathsf{Top}(Z, X)$ and $\mathsf{Top}(Z, Y)$ are isomorphic for all objects Z. This notion of gleaning information by "probing" one object with another is used extensively throughout algebraic topology where a famously useful probing object for topological spaces is the circle S^1 (and the sphere S^n, more generally).

A continuous map $S^1 \to X$ is a loop within the space X, so comparing $\mathsf{Top}(S^1, X)$ and $\mathsf{Top}(S^1, Y)$ amounts to comparing the set of all loops within X and Y. In practice, however, these are massively complicated sets. To declutter the situation, it's better to consider homotopy classes of loops, where loops aren't distinguished if one can be reshaped continuously to another. The question arises: "What are the most 'fundamental' loops in X, and do they differ from those in Y?" To simplify things further, it helps to consider only those loops that start and end at a fixed point in X. The set of such homotopy classes of loops forms a group called the *fundamental group* of X at the chosen point, which defines the object assignment of a functor $\mathsf{Top} \to \mathsf{Grp}$. And with this, the meaning of "algebraic topology" begins to come to life.

These ideas motivate a more general categorical study of pointed topological spaces and homotopy classes of maps between them. That is the goal of this chapter. Along the way, we'll encounter an interesting zoo of examples of such spaces; natural maps between them; and various adjunctions involving paths, loops, cylinders, cones, suspensions, wedges, and smashes. We open with a brief refresher in section 6.1 on homotopies and alternate ways of viewing them. In section 6.2 we motivate homotopy classes of based loops as a special case of a general construction called the fundamental groupoid. Focusing on "pointed" topological spaces in section 6.3 results in the fundamental group. We then analyze the pointed version of the product-hom adjunction, called the smash-hom adjunction, in section 6.4, and specializing yet further we obtain the suspension-loop adjunction in section 6.5. This adjunction and accompanying results on fibrations in section 6.6 provide a wonderfully short proof that the fundamental group of the circle is $\mathbb{Z}$. In section 6.6.3 we will showcase four applications of this result, and in section 6.7 we'll introduce the Seifert van Kampen theorem and use it to compute the fundamental groups of other familiar topological spaces.

6.1 Cylinder-Free Path Adjunction

Adjunctions provide us with several different but equivalent pictures of homotopies. As usual, let $I = [0, 1]$. Recall from section 1.6 that a homotopy $h \colon I \times X \to Y$ between maps f and g is a map from $I \times X \to Y$ that on one end is the map $f = h(0, -) \colon X \to Y$ and the other end is the map $g = h(1, -) \colon X \to Y$. When such a homotopy exists, f and g are said to be homotopic, and we'll write $f \simeq g$. In this setup, one usually thinks of continuously reshaping f into g over time, which is parametrized by the unit interval. Because I is locally compact and Hausdorff, there is an adjunction between the functors $I \times -$ and $(-)^I$. We call the setup

$$I \times - \colon \mathsf{Top} \rightleftarrows \mathsf{Top} \colon (-)^I$$

the *cylinder-free path adjunction* since for a space X the space $I \times X$ is the cylinder on X and X^I is the space of paths in X. These paths are described as "free" to contrast them with a *based* path space—the space of all paths beginning at a given point—which we'll consider later on.

Under the cylinder-free path adjunction, there is a bijection $\mathsf{Top}(I \times X, Y) \cong \mathsf{Top}(X, Y^I)$ for any space Y. On the left-hand side are homotopies between maps $X \to Y$; on the right-hand side are maps that associate a point in X to a path in Y. This bijection is fairly intuitive. Suppose that h is a homotopy from f to g. Then for each $x \in X$ there is a path in Y from fx to gx. Simply fix x during the homotopy $h \colon I \times X \to Y$. This path is precisely the adjunct of h evaluated at x. So a homotopy can be viewed as a function from X to the paths in Y whose value at each x is a path from fx to gx.

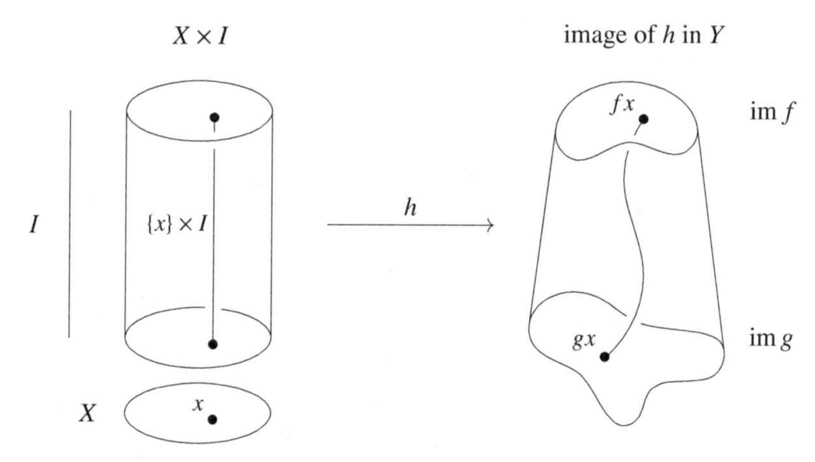

Let's consider yet another way to view homotopies. For any spaces X and Y, the compact-open topology on Y^X is splitting, so we have an injection

$$\mathsf{Top}(X \times I, Y) \to \mathsf{Top}(I, Y^X)$$

So a homotopy may also be viewed as a map $I \to Y^X$. That is, a homotopy from f to g can be viewed as a path from the point f to the point g within the space of continuous functions Y^X. If X is locally compact Hausdorff, or if we're working in the category CGWH using the k-ified product and k-ified compact-open topologies, then $Y^{X \times I} \cong (Y^X)^I$ and so every path in the mapping space Y^X defines a homotopy.

As the next theorems show, the space X, the cylinder $X \times I$, and the free path space X^I are indistinguishable in the eyes of homotopy theory. That is, these three spaces are homotopy equivalent. Homotopy equivalence was introduced in section 1.6; we'll restate it here for ease of reference.

Definition 6.1 Topological spaces X and Y are called *homotopy equivalent* if and only if there exist maps $f: X \to Y$ and $g: Y \to X$ with $fg \simeq \mathrm{id}_Y$ and $gf \simeq \mathrm{id}_X$. In this case, we write $X \simeq Y$ and f (or g) is called a *homotopy equivalence*. The category hTop is the category whose objects are spaces and whose morphisms are homotopy classes of continuous maps. So, $X \simeq Y$ if and only if $X \cong Y$ in hTop.

Theorem 6.1 The map $\pi: X^I \to X$ defined by $\gamma \mapsto \gamma 0$ and the map $i: X \to X^I$ defined by $x \mapsto c_x$, the constant path at x, are homotopy inverses.

Proof. Note that $i\pi: X^I \to X^I$ is the map that sends a path γ to $c_{\gamma 1}$, the constant path at $\gamma 1$. Let $h: X^I \times I \to X^I$ by $h(\gamma, t) = \gamma_t$ where $\gamma_t: I \to X$ is given by $\gamma_t s = \gamma(s + t - st)$. Then $h(\gamma, 0) = \gamma$ and $h(\gamma, 1) = c_{\gamma 1}$, and we see h is a homotopy between id_{X^I} and $i\pi$. Since $\pi i = \mathrm{id}_X$, there's nothing more to check. □

Theorem 6.2 The map $i: X \to X \times I$ defined by $x \mapsto (x, 1)$ and the projection $p: X \times I \to X$ are homotopy inverses.

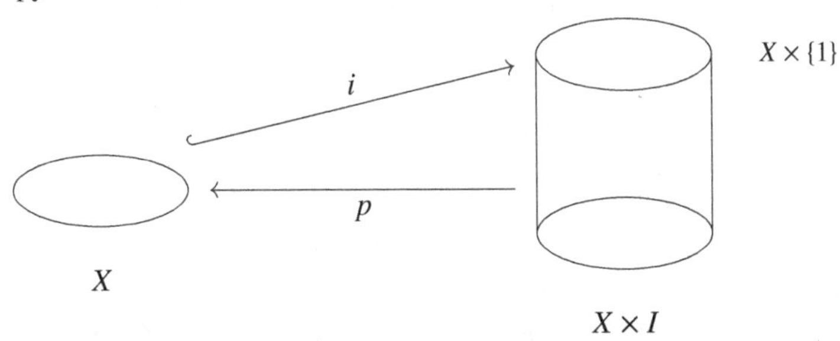

Proof. Exercise. □

In any discussion on paths and homotopy, one is prompted to think of homotopies *between* paths. Category theory provides a good setting in which to explore this. To every space, one can associate to it a category whose objects are the points in that space and whose morphisms are homotopy classes of paths between those points. This category is called the fundamental groupoid.

6.2 The Fundamental Groupoid and Fundamental Group

Let's first warm up with a few words about groupoids in general. As the name suggests, a groupoid is like a group. The connection is clear when one views groups from a categorical perspective.

A group, as noted in chapter 0, is a category with a single object where every morphism is an isomorphism. A groupoid is a slight generalization of this.

Definition 6.2 A *groupoid* is a category in which every morphism is an isomorphism.

For example, a category with two objects, each with identity morphisms, and an invertible morphism between the two is a groupoid.

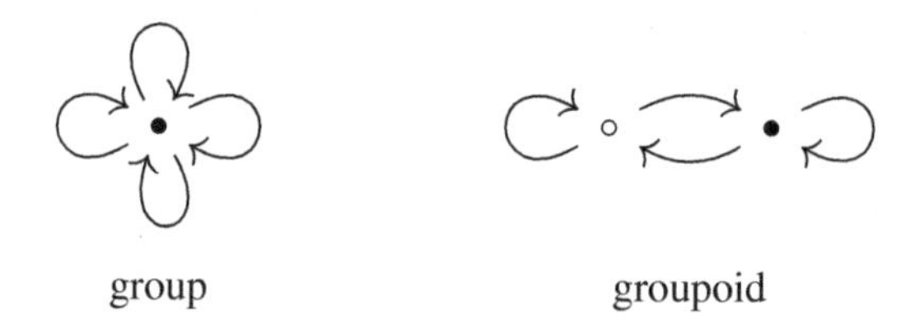

So groupoids are categories that are like groups but have possibly more than one object. For this reason, groupoids are a great way to gain intuition or inspiration when working with ideas in category theory. For example, ignoring the direction of arrows in a category results in a groupoid and hence in a structure that is very much like a group. In that sense, you might hope to "lift" theorems about groups to theorems about groupoids and then to theorems about categories. The Yoneda lemma is an example of this. It can be viewed as a lifting of *Cayley's theorem* in group theory—think about what the Yoneda lemma says when the category in question is a group.

Groupoids also appear naturally in topology when considering paths in a space. Think of a space as a category whose objects are the points in the space, and picture the morphisms to be paths between points. This doesn't quite work, though. The complication is that composition of paths isn't quite associative since the parametrization, and not just the image, is involved for paths defined as maps from the unit interval. But composition of paths is associative *up to path homotopy*. Path homotopy was defined in chapter 1. For convenience, we'll restate it here, using f and g for paths instead of α and β because we're about to think of paths as morphisms.

Definition 6.3 Two paths $f, g\colon I \to X$ from x to y in a space X are *path homotopic* if and only if there exists a homotopy $h\colon I \times I \to X$ from f to g that satisfies $h(0,t) = x$ and $h(1,t) = y$ for all t.

Under the product-hom adjunction, the homotopy may be viewed as a map $I \to X^I$ that lands in the subspace of X^I consisting of paths with fixed endpoints in X. A picture to have in mind is the cartoon below, which illustrates a simple homotopy between paths f and g. The picture can be interpreted in two equivalent ways. First, you can think of the homotopy as an extension of the paths to the shaded region. At each time t there is a path $h(-,t)$ from x to y. As t ranges from 0 to 1, one imagines f traversing the shaded region toward g. Alternatively, the homotopy can be viewed as a continuously varying family of paths from f to g. Indeed, for each point $s \in I$ there is a path from fs to gs.

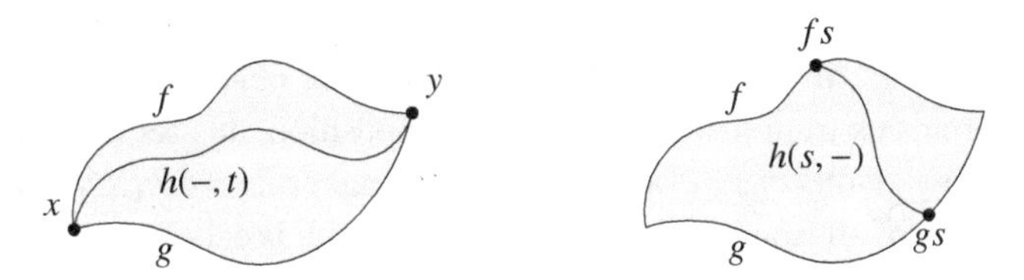

Given any two points $x, y \in X$ and a path f from one to the other, we'll be interested in the class $[f]$ of all paths from x to y that are path-homotopic to f. Points in X together with homotopy classes $[f]$ define a groupoid.

Definition 6.4 The *fundamental groupoid* $\pi_1 X$ of a space X is the category whose objects are points of X. A morphism $x \to y$ is a homotopy class of paths from x to y. Composition is given by *concatenation* of paths $[g] \circ [f] := [g \cdot f]$.

As introduced in section 2.1, if one has a path f and a path g, which begins where f ends, then their concatenation $g \cdot f$ is also a path, defined by

$$(g \cdot f)t = \begin{cases} f(2t) & 0 \le t \le \frac{1}{2} \\ g(2t-1) & \frac{1}{2} \le t \le 1 \end{cases}$$

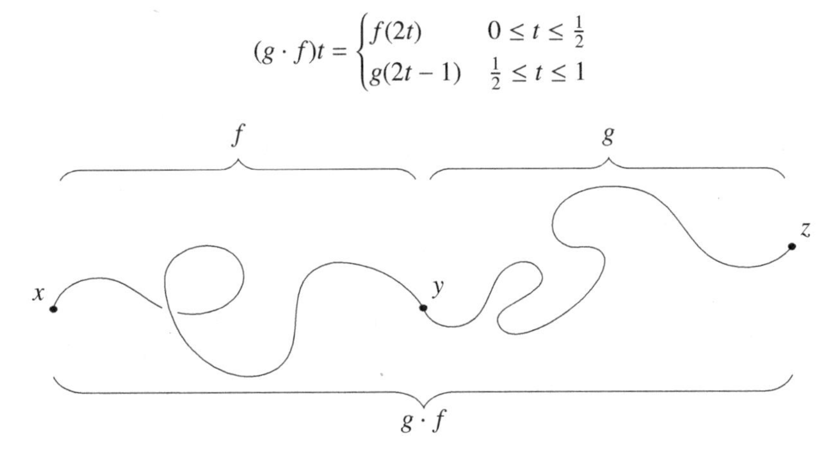

This definition provides composition in $\pi_1 X$, making $\pi_1 X$ into a category. There are a few things to verify. First, check that composition is well defined: if $f' \simeq f$ and $g' \simeq g$ then $g' \cdot f' \simeq g \cdot f$. Also check that composition of homotopy classes of paths is associative, i.e., $(h \cdot g) \cdot f \simeq h \cdot (g \cdot f)$ for any three composable paths f, g, h. Finally, check that the homotopy class of the constant path at a point x is the identity morphism id_x. One further shows that $\pi_1 X$ is a groupoid by running paths in reverse. That is, for any path f from x to y, define a path g by $gt := f(1-t)$. This is a path from y to x and it satisfies $[g][f] = \mathrm{id}_x$ and $[f][g] = \mathrm{id}_y$. So every morphism in $\pi_1 X$ is an isomorphism.

Now consider the category Grpd whose objects are small groupoids and whose morphisms are functors between groupoids. It is not difficult to check that the fundamental groupoid defines a functor $\pi_1 \colon \mathsf{Top} \to \mathsf{Grpd}$. There's no question about what π_1 does on $\mathsf{Top}(X, Y)$. The pushforward of a morphism $\alpha \in \mathsf{Top}(X, Y)$ gives a map on paths $\alpha_* : X^I \to Y^I$. One only needs to check that α_* respects homotopy equivalence.

As a general remark, in any category C one can fix an object $X \in \mathsf{C}$ and consider the set of all isomorphisms from X to itself. Under composition, this set forms a group $\operatorname{Aut} X$. In particular, fixing an object $x_0 \in X$ in the fundamental groupoid $\pi_1 X$ yields a one-object category consisting of all isomorphisms from x_0 to itself. It is called *the fundamental group of X based at the point* x_0 and is denoted by $\pi_1(X, x_0)$.

Definition 6.5 The *fundamental group of X based at the point* x_0 is the group $\pi_1(X, x_0)$ of homotopy classes of loops based at $x_0 \in X$.

As another general remark, if G is a groupoid and x is any object in G, then $\operatorname{Aut} x$ thought of as a category with one object is a full subcategory of G. (Here we momentarily denote our object x with a lowercase letter, bearing in mind points in a topological space.) The inclusion of $\operatorname{Aut} x$ into the category G defines a fully faithful functor. If G is connected, meaning that there is a morphism between any two objects, then the inclusion of $\operatorname{Aut} x$ as a subcategory of G is also essentially surjective. Therefore, for any object x of a connected groupoid, there is an equivalence of categories between the group $\operatorname{Aut} x$ and the groupoid G. So, for any path connected space X, the fundamental group is equivalent as a category to the fundamental groupoid, and moreover, for any $x_0, x_1 \in X$, the fundamental groups are isomorphic $\pi_1(X, x_0) \cong \pi_1(X, x_1)$.

Before we compute and make use of fundamental groups, we will spend the next section giving another interpretation of the fundamental group that is very good to know. Notice that a loop that begins and ends at a point x_0 is the same as a continuous function from the circle S^1 to X that sends the point $(1, 0) \in S^1$ to the point $x_0 \in X$. In this way, $\pi_1(X, x_0)$ may be viewed as homotopy classes of basepoint-preserving maps $S^1 \to X$; that is, maps that send a chosen point in S^1 to a chosen point in X. An effective, categorical way to treat basepoint-preserving maps is to consider a pair of spaces together with a point (X, x_0) as a single object and to think of basepoint-preserving maps as morphisms between these single objects.

6.3 The Categories of Pairs and Pointed Spaces

Define the *category of pairs of topological spaces* to be the category whose objects are pairs (X, A) where X is a topological space and A is a subspace of X. A morphism $f\colon (X, A) \to (Y, B)$ is a continuous function $f\colon X \to Y$ with $fA \subseteq B$. When we consider pairs for which the subset A consists of a single point, we obtain the category Top_* of pointed topological spaces. Objects in Top_* are pairs (X, x_0) where X is a topological space and x_0 is a designated point in X called the basepoint. A morphism is a continuous function $f\colon (X, x_0) \to (Y, y_0)$ with $fx_0 = y_0$. Such maps are said to *preserve* or *respect* basepoints.

Sometimes we'll write X for the pair (X, x_0) if it's understood that X has a basepoint $x_0 \in X$. In the case when X is locally compact and Hausdorff so that $\mathsf{Top}(X, Y)$ with the compact-open topology is exponential, then the set $\mathsf{Top}_*(X, Y)$ becomes a pointed space itself; it gets a topology as a subspace of $\mathsf{Top}(X, Y)$, and its basepoint is the constant map from X to the basepoint of Y.

It can be tempting to think that a choice of a basepoint in a space is only a matter of bookkeeping and that Top and Top_* may not be too different. But the categories do differ and in significant ways. For starters, colimits in Top_* are different than in Top. For example, the one-point space $*$ is both terminal and initial in Top_*, but it is not initial in Top. As we'll see in the next section, coproducts differ as well, though products do not.

Homotopies are more straightforward. A homotopy between spaces in Top_* is required to respect the basepoints: it's a map $h\colon I \times X \to Y$ such that $h(t, x_0) = y_0$ for all $t \in I$, which we'll call a *based homotopy*. And just as Top has a homotopy version hTop, so also does Top_* have a homotopy version hTop_*. This is the category whose objects are pointed topological spaces X and whose morphisms $X \to Y$ are homotopy classes of basepoint-preserving maps, the set of which is denoted by $\mathsf{hTop}_*(X, Y)$ or also by $[X, Y]_*$ or simply $[X, Y]$ if it's understood that we're working with basepoints. If X is locally compact and Hausdorff, then $[X, Y]$ is itself a pointed space. It gets a topology as a quotient of $\mathsf{Top}_*(X, Y)$, and its basepoint is the homotopy class of the constant map from X to the basepoint of Y.

We opened this chapter with a particular focus on the unit interval I and corresponding functor $\mathsf{Top}(I, -) : \mathsf{Top} \to \mathsf{Top}$. There we probed a space X with I and obtained the space of paths X^I. In the context of pointed spaces and homotopy classes of based maps, a fruitful choice of "probing space" is the sphere. By convention, the sphere S^n is a pointed space with basepoint $(1, 0, 0, \ldots, 0)$. Sometimes we'll refer to the basepoint of S^1 as 1 rather than $(1, 0)$, thinking of S^1 as the set of complex numbers z with $|z| = 1$. So one may consider the corresponding functor $\mathsf{Top}_*(S^n, -) = [S^n, -]$. When $n = 1$ this is precisely the fundamental group. That is, we have the interpretation of the fundamental group of a pointed space (X, x_0) as the set of homotopy classes of based maps from $(S^1, 1)$ to (X, x_0):

$$\pi_1(X, x_0) = [(S^1, 1), (X, x_0)]$$

This perspective has advantages. For one, it makes it clear that the fundamental group is a functor $\pi_1\colon \mathsf{Top}_* \to \mathsf{hTop}_* \to \mathsf{Set}$. It also makes it clear that it fits into a family of functors: for each $n = 0, 1, \ldots$ we have a homotopy functor $\pi_n := [S^n, -]$ from Top_* to Set called the nth *homotopy group*. What's not clear is that these functors land in Grp. The name will be justified in corollary 6.4.1 where it's shown that $\pi_n(X, x_0)$ is a group whenever $n \geq 1$.

Before moving on, we should look at the case when $n = 0$ since we've already defined a functor $\pi_0\colon \mathsf{Top} \to \mathsf{Set}$ in section 2.1.2. First, note that S^0 consists of the two points -1 and 1, and since any map $f\colon (S^0, 1) \to (X, x_0)$ must preserve basepoints, we have $1 \mapsto x_0$. So the map f is the same as a choice for $f(-1)$; that is, a point $* \to X$. Since homotopies between two such maps must also preserve basepoints, a homotopy is simply a path from one point to the other. This is consistent with our previous discussion of π_0 as the set of path components of X. If X has a basepoint, then $\pi_0(X)$ is a pointed set, the basepoint being the connected component of the basepoint of X.

So far, we've discussed some ideas in Top along with their based version in Top_*—objects, morphisms, homotopies, mapping spaces. Next, we turn to the product-hom adjuction $(X \times -) \dashv \mathsf{Top}(X, -)$ and its based version in Top_*. We already have the analogous mapping space: if X and Y have basepoints x_0 and y_0, then $\mathsf{Top}_*(X, Y)$ becomes a space as a subspace of $\mathsf{Top}(X, Y)$. The space $\mathsf{Top}_*(X, Y)$ is also based with the constant function $X \mapsto y_0$ as the basepoint. The Cartesian product $X \times Y$ also has a basepoint, (x_0, y_0). But as we'll see in the next section, the functor $X \times -$ is *not* left adjoint to $\mathsf{Top}_*(X, -)$. This motivates a new construction in Top_*, resulting in an adjunction that is the based version of the product-hom adjunction.

6.4 The Smash-Hom Adjunction

We'll begin with a more categorical discussion of the product in Top_*. To start, observe that there's a forgetful functor $U\colon \mathsf{Top}_* \to \mathsf{Top}$ that forgets basepoints. As usual, one should ask if it has a left adjoint. It does. The plus construction $+\colon \mathsf{Top} \to \mathsf{Top}_*$ is a left adjoint of U defined on objects by adding a point $X \mapsto X \coprod \{*\}$ and on morphisms $f\colon X \to Y$ by $f \mapsto \tilde{f}$ where $\tilde{f}$ extends f by sending the extra point of X to the extra point of Y. Notice that $\mathsf{Top}_*(+X, Y) \cong \mathsf{Top}(X, UY)$, so we have the adjunction

$$+\colon \mathsf{Top}_* \rightleftarrows \mathsf{Top}\colon U$$

This implies that U preserves limits. For example, the product of based spaces $(X, x_0) \times (Y, y_0)$ must be the product of the underlying topological spaces, and its basepoint is the point (x_0, y_0). Colimits, however, are different in Top_* compared with those in Top, as we hinted at earlier. In particular, the coproduct of pointed spaces X and Y is given a special name: the *wedge product*.

Definition 6.6 For pointed spaces (X, x_0) and (Y, y_0) the *wedge product* $X \vee Y$ is the quotient $X \sqcup Y/\!\sim$ where $x_0 \sim y_0$. The basepoint $*$ of $X \vee Y$ is the class $[x_0] = [y_0]$.

That is, the wedge product $X \vee Y$ is constructed by glueing X and Y together at their basepoints, and the basepoint of the new space is the single point where the two basepoints of X and Y have been identified. For example,

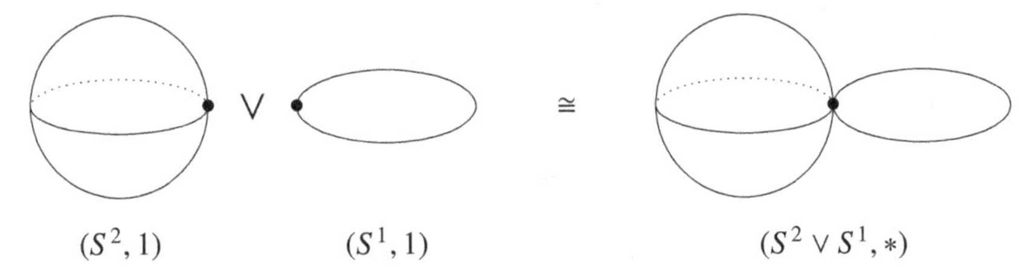

Note that there are inclusion maps

$$(X, x_0) \xrightarrow{i_X} (X \vee Y, *) \xleftarrow{i_Y} (Y, y_0)$$

which together with the wedge product satisfy the following universal property. For any pointed space (Z, z_0) and any maps $f_X\colon (X, x_0) \to (Z, z_0)$ and $f_Y\colon (Y, y_0) \to (Z, z_0)$, there exists a unique map $f\colon (X \vee Y, *) \to (Z, z_0)$ so that $f_X = fi_X$ and $f_Y = fi_Y$.

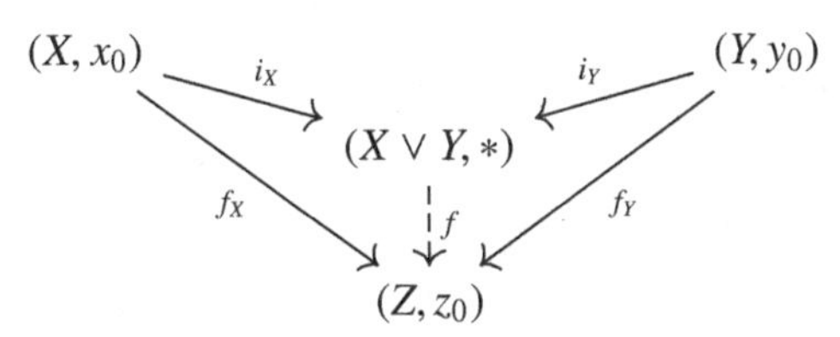

We can establish this universal property for the wedge product in Top_* by using some of the universal properties already known for familiar coproducts in Top. First observe that a pointed space (X, x_0) is the same as a space X together with a map $* \to X$. So we can view the wedge product as the pushout:

$$\begin{array}{ccc} * & \longrightarrow & Y \\ \downarrow & & \downarrow \\ X & \longrightarrow & X \vee Y \end{array}$$

The universal property for the pushout implies that for any pointed space Z, a pair of basepoint-preserving maps $X \to Z$ and $Y \to Z$ is the same as a basepoint-preserving map $X \vee Y \to Z$. In other words, the wedge product is the coproduct in Top_*. We'll use this coproduct to give a refined version of the product-hom adjunction for based spaces.

To begin, consider the identification $\mathsf{Top}(X\times Y, Z) \cong \mathsf{Top}(Y, Z^X)$. In the context of pointed spaces, there are a couple of things to consider. First, if x_0 and z_0 are basepoints in Z and X respectively, then Z^X has a basepoint given by the constant function $f_0\colon X \to z_0$. A map $f\colon Y \to Z^X$ on the right-hand side must preserve basepoints, meaning that it must satisfy $(fy_0)x = z_0$ for all $x \in X$. Additionally, for any $y \in Y$ the resulting map $fy\colon X \to Z$ must also preserve basepoints. That is, $(fy)x_0 = z_0$ for all $y \in Y$. Therefore, if the adjoint of a map $f\colon X \times Y \to Z$ is a basepoint preserving map $Y \to Z^X$, then f must be constant on $(\{x_0\} \times Y) \cup (X \times \{y_0\})$, sending it to z_0. This motivates the definition of the *smash product* of topological spaces.

Definition 6.7 The *smash product* of two pointed spaces (X, x_0) and (Y, y_0) is defined to be the quotient space

$$X \wedge Y \quad := \quad X \times Y/X \vee Y$$

where $X \vee Y$ is identified with the subspace $(\{x_0\} \times Y) \cup (X \times \{y_0\})$. It has (x_0, y_0) as a basepoint.

When X is locally compact and Hausdorff, the smash product is the quotient of the product by the minimal relation that ensures that there is a bijection of sets $\mathsf{Top}_*(X \wedge Y, Z) \cong \mathsf{Top}_*(Y, Z^X)$. So it is not surprising that the naturality of the product-hom adjunction in Top descends to yield the *smash-hom adjunction* among pointed spaces:

$$X \wedge -\colon \mathsf{Top}_* \rightleftarrows \mathsf{Top}_*\colon (-)^X$$

As we'll see next, an important case of the smash-hom adjuntion arises when X is taken to be the circle.

6.5 The Suspension-Loop Adjunction

There is a special name given to the smash product of the circle with a pointed space X—the *reduced suspension* of X. One can also smash X with the unit interval to obtain the *reduced cone* over X.

Definition 6.8 For a pointed space (X, x_0), the *reduced cone* CX and the *reduced suspension* are given by

$$CX := X \wedge I \quad \text{and} \quad \Sigma X := X \wedge S^1$$

More concretely,

$$\begin{aligned} CX &= X \times I/\!\sim \quad \text{where} \quad (x,1) \sim (x',1) \\ & \qquad\qquad\qquad\qquad\;\; (x_0,t) \sim (x_0,s) \end{aligned}$$

$$\begin{aligned} \Sigma X &= X \times I/\!\sim \quad \text{where} \quad (x,0) \sim (x',0) \\ & \qquad\qquad\qquad\qquad\;\; (x,1) \sim (x',1) \\ & \qquad\qquad\qquad\qquad\;\; (x_0,t) \sim (x_0,s) \end{aligned}$$

for all $x, x' \in X$ and $s, t \in I$, and the basepoints of CX and ΣX are the classes $[x_0] = \{x_0\} \times I$. Simple sketches show that they look like quotients of (i) a cone drawn up from X to a point and (ii) a copy of X suspended between two points—one above, one below—as though by rigging lines. See figure 6.1.

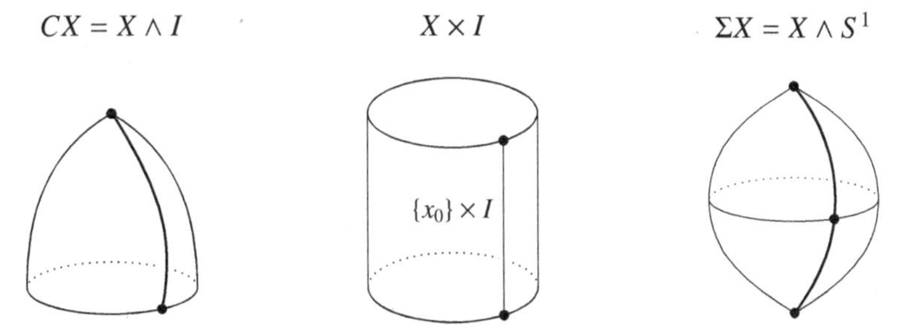

Figure 6.1 Sketches of the reduced cone (left) and reduced suspension (right), where points along the bold lines are identified.

Notice the identifications in the quotient of ΣX are consistent with those in definition 6.7—here we are using the fact that S^1 is obtained from I by identifying the endpoints 0 and 1. Now, if X does not have a basepoint, then we have the analogous "unreduced" constructions.

Definition 6.9 Let X be a topological space. The *cone* CX and *suspension* SX are defined to be

$$CX := X \times I/\!\sim \quad \text{where} \quad (x,1) \sim (x',1)$$

$$\begin{aligned} SX &:= X \times I/\!\sim \quad \text{where} \quad (x,0) \sim (x',0) \\ & \qquad\qquad\qquad\qquad\;\;\; (x,1) \sim (x',1) \end{aligned}$$

While the terminology may be new, suspensions arise in a familiar context: compactifications. The one-point compactification of $\mathbb{R}$, for example, is the circle S^1, and the one-point compactification of $\mathbb{R} \times \mathbb{R}$ is the sphere S^2. But another compactification of $\mathbb{R} \times \mathbb{R}$ is the torus $S^1 \times S^1$. How are these related? By the main property of the one-point compactification, S^2 is the quotient of $S^1 \times S^1$ by the extra points $(1 \times S^1) \cup (S^1 \times 1)$. That is, $S^1 \wedge S^1 \cong S^2$:

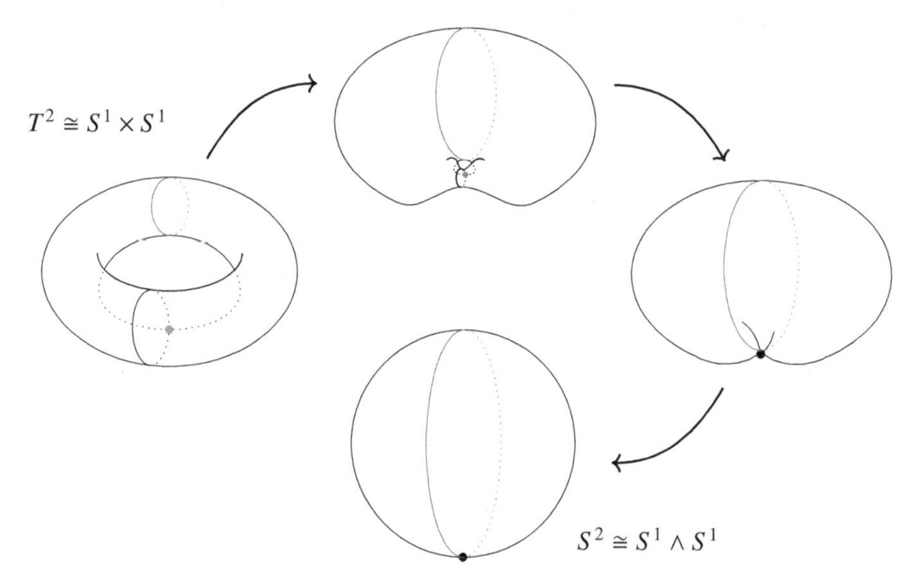

This argument works in general.

Theorem 6.3 Suppose X^* and Y^* are one-point compactifications of spaces X and Y. Then

$$X^* \wedge Y^* \quad \cong \quad (X \times Y)^*$$

where the extra points at infinity are the basepoints.

As a result, suspensions give a simple relationship between S^n and S^{n+1}.

Corollary 6.3.1 $\Sigma S^n \cong S^{n+1}$ for $n \geq 0$.

Notice that $S^1 \wedge -$ and $(-)^{S^1}$ are both functors from Top_* to Top_*, called the *reduced suspension* Σ and the *based loop functor* Ω, respectively. This gives rise to the *suspension-loop adjunction.*

$$\Sigma \colon \mathsf{Top}_* \rightleftarrows \mathsf{Top}_* : \Omega$$

The correspondence $\mathsf{Top}_*(\Sigma X, Y) \cong \mathsf{Top}_*(X, \Omega Y)$ is understood as follows. Suppose we have a map $f : \Sigma X \to Y$, and for any point $x \in X$, consider the subspace $\{x\} \times I$ of the cylinder $X \times I$. After forming the quotient ΣX, this space becomes $\{x\} \times S^1$, which is then mapped to Y via f. The assignment sending x to this loop is the adjunct of f. In particular, f must map the basepoint $*$ of ΣX to the basepoint y_0 of Y. This gives a map from $*$ to the constant loop at y_0. If we further pass to (basepoint-preserving) homotopy classes of morphisms, then we obtain for every pair of pointed spaces X and Y,

$$[\Sigma X, Y] \cong [X, \Omega Y]$$

which has important consequences.

Theorem 6.4 Let X be a pointed space. Then $\pi_n X \cong \pi_{n-1}\Omega X$ for each $n \geq 1$.

Proof. By corollary 6.3.1 and the suspension-loop adjunction,

$$\begin{aligned} \pi_n X &= [S^n, X] \\ &\cong [\Sigma S^{n-1}, X] \\ &\cong [S^{n-1}, \Omega X] \\ &= \pi_{n-1}\Omega X \end{aligned}$$

□

Since this construction is repeatable, we find that $\pi_n X \simeq \pi_1 \Omega^{n-1} X$. We already know that the fundamental group of any space is a group, and so we now know that the higher homotopy groups are also a group.

Corollary 6.4.1 Let (X, x_0) be a pointed space. Then $\pi_n(X, x_0)$ is a group for each $n \geq 1$.

In addition, it turns out that $\pi_n X$ is abelian if $n \geq 2$. We'll hint at the proof in theorem 6.6. Our next goal is to compute the fundamental group of some familiar spaces, starting with the circle. One would think the circle is a rather benign space, but to compute its fundamental group, we will use some new machinery—fibrations.

6.6 Fibrations and Based Path Spaces

Often in mathematics, one is interested in organizing similar objects into families. Usually, this is formalized as follows: one has a "total space" E that maps onto a "base" space B. The objects being organized into families are the fibers of the map $E \to B$. With this in mind, let us describe a situation that arises in homotopy theory. We have a map of topological spaces $p\colon E \to B$, and there is another space X that lies inside E in some way—say, $g\colon X \to E$—and which also lies in the base space B as $pg\colon X \to B$. Now suppose X lies within B as the initial part of a homotopy $h\colon X \times I \to B$. In this setup, one views the copy of X within E as being the first step in "lifting" the homotopy from B up to E. A natural question is: *Can we finish the task?* That is, can the rest of the homotopy in B be lifted to E? If the answer is "yes," then the map p is called a *fibration*.

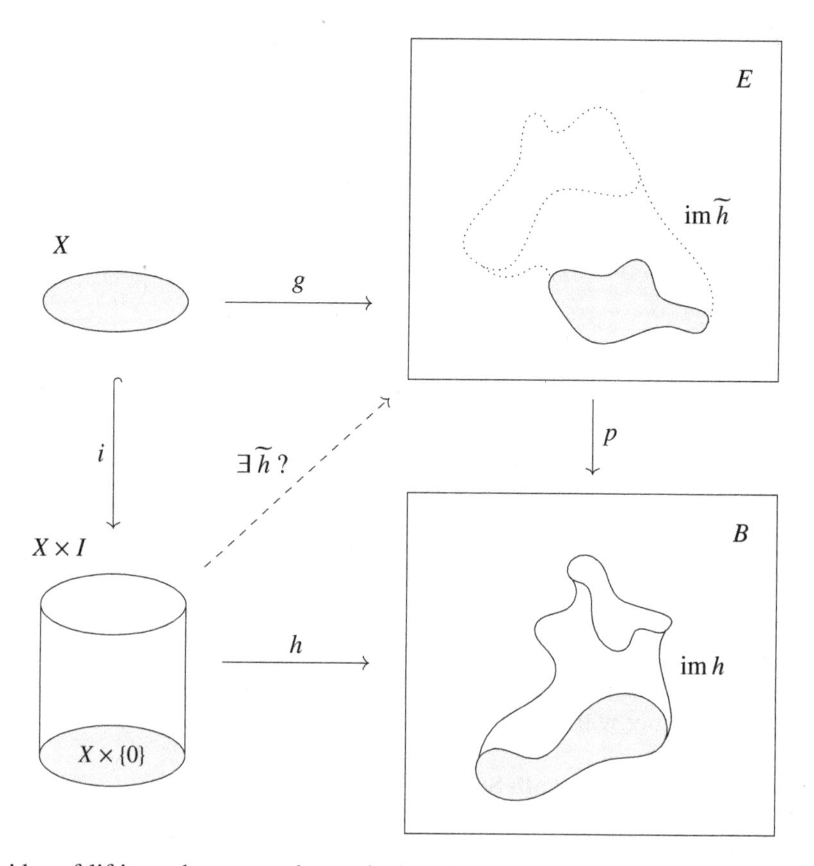

This idea of lifting a homotopy has a dual notion: extending a homotopy. Start with a homotopy $A \times I \to X$, and suppose A sits inside another space Y by way of a map $i\colon A \to Y$. It's natural to wonder if the homotopy extends to Y. In particular, if a map $g\colon Y \cong Y \times \{0\} \to X$ already exists and is thought of as the first step of an extension, then one might hope that a full extension exists. If it does, then the map i is called a *cofibration*.

Definition 6.10 A map $p\colon E \to B$ is called a *fibration* if and only if for any maps h and g making the outer square commute, there exists $\widetilde{h}\colon Y \to X^I$ so that the whole diagram commutes:

$$\begin{array}{ccc} X & \xrightarrow{g} & E \\ {\scriptstyle i}\downarrow & \overset{\widetilde{h}}{\nearrow} & \downarrow{\scriptstyle p} \\ X \times I & \xrightarrow{h} & B \end{array}$$

Here i is the map sending each $x \in X$ to $(x, 0)$. Often, E is referred to as the *total space* of the fibration while B is called the *base space* of the fibration.

Dually, a map $i\colon A \to Y$ is a *cofibration* if for any maps h and g making the outer square commute, there exists $\widetilde{h}\colon Y \to X^I$ so that the whole diagram commutes:

$$\begin{array}{ccc} A & \xrightarrow{h} & X^I \\ {\scriptstyle i}\downarrow & \overset{\widetilde{h}}{\nearrow} & \downarrow{\scriptstyle e} \\ Y & \xrightarrow{g} & X \end{array}$$

Here the map e is evaluation at 0. It sends a path γ to its starting point $\gamma(0)$.

These are challenging definitions. Here's the takeaway. Fibrations are maps into B with the property that, *if a homotopy in B lifts to the total space at a point, then it lifts completely.* Cofibrations are the maps out of space A where, *if a homotopy in A extends at a point, then it extends completely.* This follows quickly from a consideration of what commutativity means in these diagrams:

$$\begin{array}{ccc} X & \xrightarrow{g} & E \\ {\scriptstyle i}\downarrow & & \downarrow{\scriptstyle p} \\ X \times I & \xrightarrow{h} & B \end{array} \qquad\Bigg|\qquad \begin{array}{ccc} A & \xrightarrow{h} & X^I \\ {\scriptstyle i}\downarrow & & \downarrow{\scriptstyle e} \\ Y & \xrightarrow{g} & X \end{array}$$

For the left diagram to commute, the map $h(-,0)\colon X \times \{0\} \to B$, which is hi up to the isomorphism $X \cong X \times \{0\}$ and must be equal to pg, which is to say g is lift of the homotopy h at the point 0. Now let's turn to the homotopy extension diagram at the right. Recall the adjunction $\mathsf{Top}(A \times I, X) \cong \mathsf{Top}(A, X^I)$. So h in the diagram corresponds to a homotopy $A \times I \to X$. And, up to the isomorphism $X \cong X \times \{0\}$, the map g between $Y \to X^{\{0\}}$ extends the homotopy h along i at the point 0.

We've referred to fibrations and cofibrations as "dual" notions, but know that some care must be taken to properly dualize concepts in mathematics—there's usually more going on under the hood. But here's the simple idea: (co)fibrations are maps for which (extensions) lifts exist as soon as we have them for a single point. These ideas respectively are called the *homotopy lifting property* and *homotopy extension property*, and you can generally think of a (co)fibration as a map with the homotopy (extension) lifting property for all spaces.

We'll give some examples below, but first, it's good to know that the lifting/extension properties enjoyed by (co)fibrations are especially potent in a discussion of homotopy theory. From the homotopical viewpoint, any continuous function *is* either a fibration or a cofibration. Let's say a few brief words about this.

6.6.1 Mapping Path Space and Mapping Cylinder

Any continuous function factors as a homotopy equivalance followed by a fibration. Dually, any continuous function factors as a cofibration followed by a homotopy equivalence. In other words, any map can be replaced by a fibration or a cofibration at will, "up to homotopy." Even better, this a constructive statement. Given any map $f\colon X \to Y$, we can explicitly construct a homotopy equivalance and fibration that realizes this factorization, and it is similar for cofibrations. The homotopy equivalences appearing in these statements involve two topological spaces associated to f: its *mapping path space* and its *mapping cylinder.*

Definition 6.11 The *mapping path space* P_f of a map f is the pullback:

$$\begin{array}{ccc} P_f & \longrightarrow & X \\ \downarrow & \lrcorner & \downarrow f \\ \mathcal{P}Y & \longrightarrow & Y \end{array}$$

In other words, P_f consists of pairs $(x, \gamma) \in X \times \mathcal{P}Y$ where $fx = \gamma 1$. Sometimes P_f is also called the *mapping cocylinder*.

A nice consequence is that the assignment $x \mapsto (x, c_{fx})$ defines a homotopy equivalence from X to P_f. (Its homotopy inverse is simply projection onto the first factor.) And the map $P_f \to Y$ sending the pair (x, γ) to $\gamma 1$ is a fibration. The idea behind the proof is similar to that in example 6.1. Then, as claimed, we have a factorization of f as a homotopy equivalence followed by a fibration:

$$X \xrightarrow{\simeq} P_f \twoheadrightarrow Y \qquad (\text{composite } f)$$

Definition 6.12 The *mapping cylinder* M_f *of* f is the pushout

$$\begin{array}{ccc} X & \longrightarrow & X \times I \\ f\downarrow & \ulcorner & \downarrow \\ Y & \longrightarrow & M_f \end{array}$$

where $X \to X \times I$ is the map $x \mapsto (x, 0)$. In other words, M_f is the quotient of the disjoint union of Y and $X \times I$ obtained by identifying fx and $(x, 0)$ for each $x \in X$. One imagines M_f as a "cylinder" whose top is a copy of X and whose base is fX, which sits inside Y.

It can be shown that the map $X \to M_f$ defined by $x \mapsto [(x, 1)]$ is a cofibration, that Y is homotopy equivalent to M_f, and that any map $f\colon X \to Y$ factors through this cofibration and equivalence,

$$X \longrightarrow M_f \xrightarrow{\simeq} Y \qquad (\text{composite } f)$$

The fact that any map is, up to homotopy, either a fibration or a cofibration composed with a homotopy equivalence suggests that the triad—fibration, cofibration, and homotopy equivalence—is intrinsically useful in an exploration of homotopy theory. Indeed, it propels one to the study of *model category theory*. A *model category* is a complete and cocomplete category with three classes of morphisms called fibrations, cofibrations, and weak equivalences that satisfy certain conditions. Or, more succinctly, it's a category in which one can "do homotopy theory." As you might hope, Top with homotopy equivalences and (co)fibrations as given in definition 6.10 provides a prime example of a model category (see Strøm (1972)), but it's not the only model structure on Top, nor is Top the only category with a model structure. We mention these ideas simply to whet the appetite. A more categorical discussion of (co)fibrations, equivalences, model categories, and general categorical homotopy theory may be found in Riehl (2014).

Let's now return to the task at hand—examples.

6.6.2 Examples and Results

To give a first example of a fibration, we'll introduce a new mapping space associated to a pointed space X.

Definition 6.13 The mapping space $\mathcal{P}X = \mathsf{Top}_*((I,0),(X,x_0))$ is called the *based path space* of X.

So points in $\mathcal{P}X$ are paths that start at x_0 and end at some $x \in X$. This space of based paths is itself a pointed space with the constant path $c_{x_0}\colon I \mapsto x_0$ serving as the basepoint.

You'll have noticed we view the interval I as a pointed space in two ways: either with basepoint 0 or with basepoint 1. When constructing the reduced cone CX on a pointed space X as $CX = X \wedge I$, we regard the basepoint of I to be 1. We like our cones to be right-side up, not upside down. On the other hand, when constructing the based path space $\mathcal{P}X$ on a pointed space X as $\mathsf{Top}_*(I,X)$, we regard the basepoint of I to be 0. We like the basepoint of X to be the beginning—rather than the end—of a path.

But the endpoints of paths are of interest. There is a map $p\colon \mathcal{P}X \to X$ which sends a path γ to its endpoint $\gamma 1 \in X$. This map provides a nice connection between $\mathcal{P}X$ and another important mapping space: the fiber $p^{-1}x_0$ consisting of all loops at x_0. That is, $p^{-1}x_0 = \Omega X$, a situation typically illustrated as a diagram:

$$\begin{array}{ccc} \Omega X & \longrightarrow & \mathcal{P}X \\ & & \downarrow{\scriptstyle p} \\ & & X \end{array}$$

There are other ways in which p is a particularly nice kind of map. It sends c_{x_0} to x_0, so it is basepoint preserving. Moreover, it is a fibration.

Example 6.1 For any based space X, the map $p\colon \mathcal{P}X \to X$ sending a path γ to its endpoint $\gamma(1)$ is a fibration. Suppose we have the commuting square

$$\begin{array}{ccc} Z & \xrightarrow{g} & \mathcal{P}X \\ \downarrow & & \downarrow p \\ Z \times I & \xrightarrow{h} & X \end{array}$$

where Z is any pointed space. Notice that for a fixed $z \in Z$, gz is a path in X ending at the point $h(z, 0)$, which, with z still fixed, is the starting point of the path h_z. So define $\widetilde{h}\colon Z \times I \to \mathcal{P}X$ to be the parameterization of the concatenation $h_z \cdot gz$ given by:

$$\widetilde{h}(z,t)s = \begin{cases} gz(s(1+t)) & \text{if } 0 \leq s \leq \frac{1}{1+t} \\ h(z, s(1+t)-1) & \text{if } \frac{1}{1+t} \leq s \leq 1. \end{cases}$$

One can check that $\widetilde{h}$ preserves basepoints and commutes with the diagram.

While on the topic of based path spaces, here is a good property to know about.

Proposition 6.1 $\mathcal{P}X$ is contractible.

Proof. Let $*$ denote the one-point space. The composition $* \to \mathcal{P}X \to *$ is equal to id_*, so to prove $\mathcal{P}X$ is homotopy equivalent to $*$, we need only show the composition $\mathcal{P}X \to * \to \mathcal{P}X$ that sends a path γ to c_{x_0} is homotopic to $\mathrm{id}_{\mathcal{P}X}$.

Define $h\colon \mathcal{P}X \times I \to \mathcal{P}X$ by $h(\gamma, t) = \gamma_t$, where $\gamma_t\colon I \to X$ is the path $\gamma_t(s) = \gamma(s + (1-s)t)$. Then h is the identity on $\mathcal{P}X$ at $t = 0$, and h is the constant map $\gamma \mapsto c_{\gamma 1}$ at $t = 1$. Further, h is basepoint-preserving since $h(c_{x_0}, t) = c_{x_0}$ for all t. □

Here is another important example of a fibration.

Example 6.2 The map p from $\mathbb{R}$ to S^1 given by $y \mapsto e^{2\pi i y}$ is a fibration with fiber $\mathbb{Z}$. That is, if the following diagram commutes

$$\begin{array}{ccc} X & \xrightarrow{g} & \mathbb{R} \\ \downarrow & & \downarrow p \\ X \times I & \xrightarrow{h} & S^1 \end{array}$$

then there is a lift of the homotopy h through p.

The key here is that p is a local homeomorphism: for every $y \in \mathbb{R}$ there is an open neighborhood of y that maps homeomorphically onto its image—for instance, the interval $(y - \frac{1}{2}, y + \frac{1}{2})$ would work. So there is an open cover $\mathcal{U}$ of S^1 so that for each $U \in \mathcal{U}$, the inverse image $p^{-1}U$ is a collection of disjoint open sets in $\mathbb{R}$, each of which is homeomorphic to U. Because h is continuous, its inverse image $h^{-1}\mathcal{U}$ is an open cover of $X \times I$. Further, on each of the compact subspaces $\{x\} \times I$ there is a finite subcover, say $\mathcal{V}_x \subseteq h^{-1}\mathcal{U}$.

Lifting h inductively along any one of the $\mathcal{V}_x$ isn't hard. We need two observations. First, if a lift $g_x \colon S \to \mathbb{R}$ is defined for a nonempty subset $S \subseteq h^{-1}U \in \mathcal{V}_x$, then g_xS is contained entirely in one of the disjoint homeomorphic copies of U in $p^{-1}U$—call it U_S. Second, p restricts to a homeomorphism $p_S : U_S \to U$. We can thus extend the domain of such a g_x to include all of $h^{-1}U$ by declaring $g_x := p_S^{-1}h$ on $h^{-1}U$. Diagrammatically, we have:

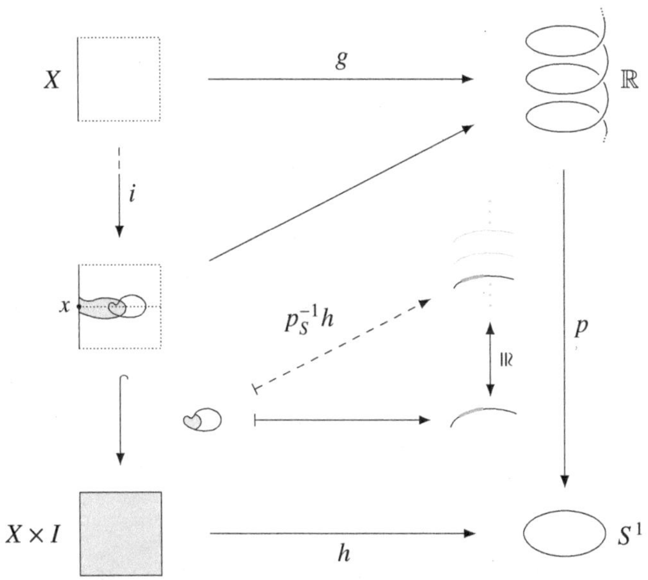

Inducting through the finite subcover (using the given g as the base step) we define for each $x \in X$ a lift g_x whose domain includes the open cover $\mathcal{V}_x$ of $\{x\} \times I$:

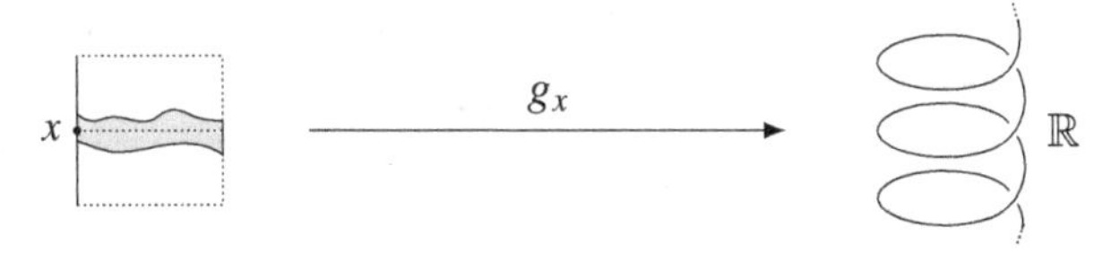

It's possible that three or more sets in the cover $\mathcal{V}_x$ overlap, in which case there is a choice of inductions extending g to a g_x. However, all such inductions must agree on any overlaps precisely because p is a local homeomorphism, so the g_x are well defined.

This same observation guarantees that any two lifts g_x and $g_{x'}$ must be equal on the intersection of their domains. Therefore the g_x assemble to uniquely determine a map

$\widetilde{h}\colon X \times I \to \mathbb{R}$ such that

$$\begin{array}{ccc} X & \xrightarrow{\;g\;} & \mathbb{R} \\ {\scriptstyle i}\big\downarrow & \overset{\widetilde{h}}{\nearrow} & \big\downarrow{\scriptstyle p} \\ X \times I & \xrightarrow{\;h\;} & S^1 \end{array}$$

commutes. In other words, the exponential map p is a fibration.

These examples give two different fibrations whose base space is the circle.

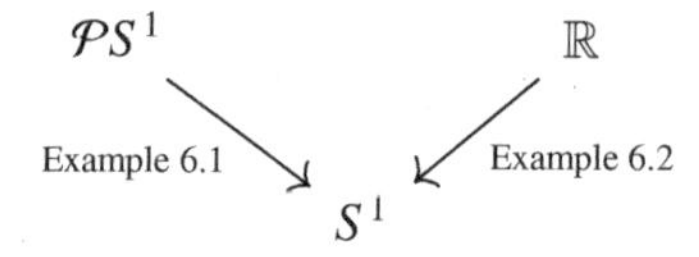

Naturally one wonders if this triangle can be completed. Are $\mathbb{R}$ and $\mathcal{P}S^1$ related? Yes, both spaces are contractible, so there is a homotopy equivalence between them. As the next theorem shows, these fibrations must therefore have homotopy equivalent fibers.

Theorem 6.5 Suppose p and q are fibrations with base space B and f is a map of total spaces causing the diagram to commute:

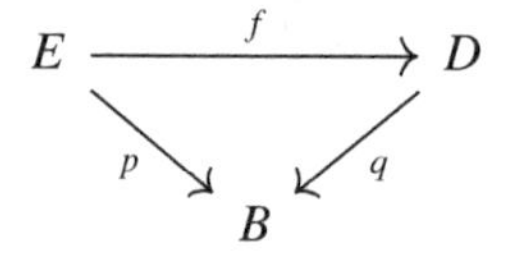

If f is a homotopy equivalence, then f induces a homotopy equivalence between fibers.

Commutativity of the triangle implies $fp^{-1}b \subseteq q^{-1}b$ for all $b \in B$, which is to say that f is a *fiber-preserving map.* But its homotopy inverse $f'\colon D \to E$ may not be fiber preserving, and the homotopies witnessing $ff' \simeq \mathrm{id}_D$ and $f'f \simeq \mathrm{id}_E$ may not be comprised of fiber-preserving maps. However, if one can replace f' with a fiber-preserving homotopy equivalent map g that satisfies $fg \simeq \mathrm{id}_D$ and $gf \simeq \mathrm{id}_E$, where each homotopy is comprised of fiber-preserving maps, then for each $b \in B$ the map f can restrict to a homotopy equivalence between fibers, $p^{-1}b \simeq q^{-1}b$. The theorem thus rests on producing such a map g.

Proof. By assumption, there is a homotopy h' from ff' to id_D. Postcomposing it with the fibration q gives homotopy $h\colon D \times I \to B$ from pf' to q, and the outer square commutes by commutativity of the triangle:

$$\begin{array}{ccc} D & \xrightarrow{\;f'\;} & E \\ \big\downarrow & \overset{\widetilde{h}}{\nearrow} & \big\downarrow{\scriptstyle p} \\ D \times I & \xrightarrow{\;h\;} & B \end{array}$$

Since p is a fibration, there is a lift $\widetilde{h}$ with $\widetilde{h}(-,0) = f'$. We claim that $g := \widetilde{h}(-,1)$ is the desired map. First note that g is fiber preserving by commutativity of the previous diagram. Moreover, fg and id_D are homotopic by $k\colon D \times I \to D$ defined by

$$k(d,t) = \begin{cases} f\widetilde{h}(d, 1-2t), & \text{if } 0 \le t \le \frac{1}{2} \\ h'(d, 2t-1), & \text{if } \frac{1}{2} \le t \le 1 \end{cases}$$

The map $k(-,t)$, however, may not be fiber preserving for each $t \in I$. To work around this, one can define a homotopy $M\colon D \times I \times I \to D$ from the homotopy qk to a homotopy between q and itself so that the square commutes:

$$\begin{array}{ccc} D \times I & \xrightarrow{k} & D \\ \downarrow & \overset{L}{\nearrow} & \downarrow q \\ D \times I \times I & \xrightarrow{M} & B \end{array}$$

Since q is a fibration, M lifts to a homotopy L. Since L fits into the diagram, it provides the desired fiber-preserving homotopy from fg to id_D:

$$fg = k(-,0) = L(-,0,0) \simeq L(-,1,0) = k(-,1) = \mathrm{id}_D$$

A similar story shows $gf \simeq \mathrm{id}_E$. Here's the sketch. First use the fibration p together with the homotopy witnessing $f'f \simeq \mathrm{id}_E$ to get a homotopy $E \times I \to B$ that fits into the square

$$\begin{array}{ccc} E & \xrightarrow{f} & D \\ \downarrow & \nearrow & \downarrow q \\ E \times I & \longrightarrow & B \end{array}$$

The lift of the homotopy defines a map $\bar{g}\colon E \to D$. One can then show $g\bar{g} \simeq \mathrm{id}_E$ through a fiber-preserving homotopy, from which it follows that $\bar{g} = \mathrm{id}_D\,\bar{g} \simeq fg\bar{g} \simeq f$ and so $gf \simeq g\bar{g} \simeq \mathrm{id}_E$. □

Right away, we obtain several important corollaries.

Corollary 6.5.1 The loop space of the circle ΩS^1 is homotopy equivalent to $\mathbb{Z}$

Proof. Both $\mathbb{R} \to S^1$ and $\mathcal{P}S^1 \to S^1$ are fibrations by earlier examples. Moreover, both $\mathcal{P}S^1$ and $\mathbb{R}$ are contractible (proposition 6.1 and example 1.21). There is thus a homotopy equivalence between them that commutes with the fibrations

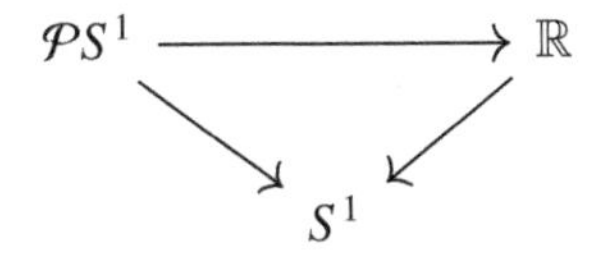

By theorem 6.5 it induces a homotopy equivalence between the fibers ΩS^1 and $\mathbb{Z}$ of $\mathcal{P}S^1 \to S^1$ and $\mathbb{R} \to S^1$, respectively. □

In fact, this idea holds in greater generality.

Corollary 6.5.2 Let $p\colon E \to B$ be a fibration with fiber F. If E is contractible, then F is homotopy equivalent to the loop space ΩB.

Immediately, we obtain the next result.

Corollary 6.5.3 The fundamental group of S^1 is isomorphic to $\mathbb{Z}$.

Proof. By the previous corollary, $\Omega S^1 \simeq \mathbb{Z}$ which implies

$$\pi_0 \Omega S^1 \cong \pi_0 \mathbb{Z}$$

The left-hand side is $\pi_1 S^1$ by theorem 6.4. The right-hand side is the set of path components of $\mathbb{Z}$, which is simply $\mathbb{Z}$. □

Now, you might be concerned that $\pi_0 \mathbb{Z}$ is merely a set with no additional structure since, after all, the corollary to theorem 6.4 held only in the case when $n \geq 1$. But by the homotopy invariance of π_0, it follows that $\pi_0 \mathbb{Z}$ is isomorphic to $\pi_0 \Omega S^1 = [S^0, \Omega S^1]$, and the latter, being a set of (homotopy classes of) maps into a group, is itself a group. So we are assured that $\pi_0 \mathbb{Z} \cong \mathbb{Z}$ is indeed a group.

Another important consequence is the following.

Corollary 6.5.4 The nth homotopy group of the circle is trivial for $n \geq 2$.

Proof. If $n \geq 2$, then

$$\pi_n S^1 = \pi_{n-1} \Omega S^1 = \pi_{n-1} \mathbb{Z} = [S^{n-1}, \mathbb{Z}]$$

The result follows since S^{n-1} is connected for $n > 1$ so any basepoint-preserving map from it to $\mathbb{Z}$ must be constant to the basepoint of $\mathbb{Z}$. □

While on the topic of nth homotopy groups, here is the result promised earlier.

Theorem 6.6 For any space X the nth homotopy group $\pi_n X$ is abelian for $n \geq 2$.

Proof. The proof is left to exercise 6.4 at the end of the chapter. The picture below gives a hint when $n = 2$.

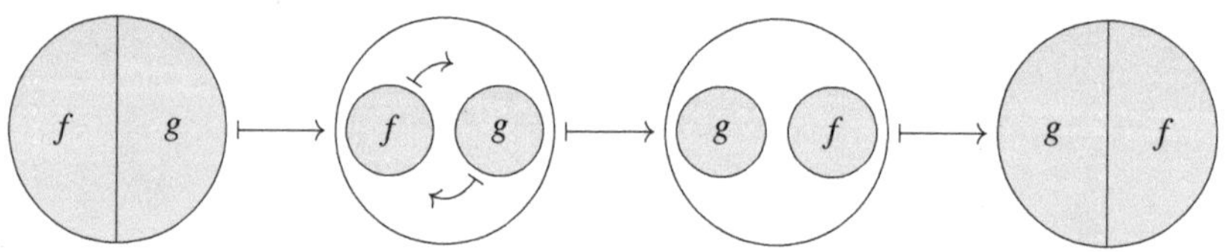

□

Now that we've computed the fundamental group of the circle, we'll present a few applications of the result. Afterward, we'll turn our attention to the fundamental groups of other familiar spaces. As we'll see then, a theorem of Seifert and van Kampen provides a methodical way of doing so.

6.6.3 Applications of $\pi_1 S^1$

The isomorphism $\pi_1 S^1 \cong \mathbb{Z}$ leads to some nice results, some of which we showcase here. To start, recall that in section 2.1 we proved that every map from the closed interval $[-1, 1]$ to itself must have a fixed point. Here is the two-dimensional analog.

Brouwer's Fixed Point Theorem Every map $D^2 \to D^2$ has a fixed point.

Proof. Suppose $f\colon D^2 \to D^2$ is a map with no fixed points. Then for any $x \in D^2$, there is a unique ray starting at fx and passing through x.

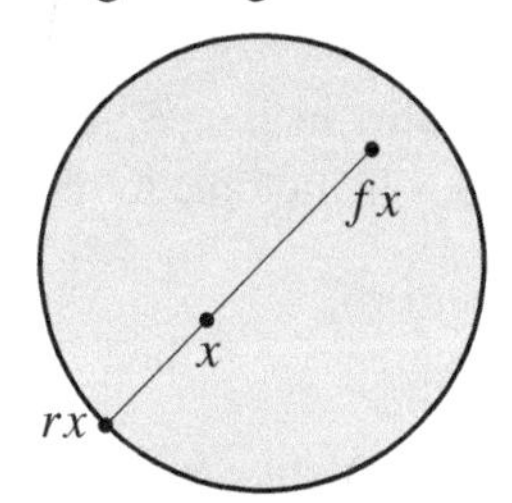

Let rx denote the point where this ray intersects the boundary of the disc $S^1 = \partial D^2$, and observe that r is a continuous map which satisfies $r(x) = x$ for all $x \in S^1$. Thus $ri = \mathrm{id}_{S^1}$ where $i\colon S^1 \hookrightarrow D^2$ is the inclusion. Choosing a basepoint in S^1 and applying π_1, we obtain

$$\pi_1(S^1, 1) \xrightarrow{\pi_1 i} \pi_1(D^2, 1) \xrightarrow{\pi_1 r} \pi_1(S^1, 1) \qquad \pi_1(id_{S^1}) = id_{\pi_1(S^1,1)}$$

But this is impossible since $\pi_1 D^2 \cong 0$ while $\pi_1 S^1 \cong \mathbb{Z}$. Emphatically, the identity does not factor through the constant map at 0. □

The following result from linear algebra is a corollary.

Perron-Frobenius Theorem Every 3×3 matrix with positive entries has a positive eigenvalue.

Proof. Let Δ^2 denote the subset of $\mathbb{R}^3$ consisting of all points (x, y, z) satisfying $x+y+z = 1$, where each coordinate lies in the interval $[0, 1]$. That is, Δ^2 is the face opposite the origin of the unit tetrahedron in the first quadrant of $\mathbb{R}^3$. Now if A is any 3×3 matrix with real positive entries, define a linear map B by

$$Bv = \frac{1}{\lambda_v} Av$$

where λ_v is the sum of the coordinates of the vector Av. Then B is a linear transformation from Δ^2 to itself. And since Δ^2 and the disc D^2 are homeomorphic, B must have a fixed point. Thus there is a vector w so that $w = Bw$ and so $Aw = \lambda_w w$ where, by assumption, λ_w must be positive. □

We have two more applications of $\pi_1 S^1$ to share. Both require the following definition.

Definition 6.14 Let $f\colon (S^1, 1) \to (S^1, f1)$. Choose a path $f1 \to 1$, which defines an isomorphism $\pi_1(S^1, f1) \simeq \pi_1(S^1, 1)$. Then $\pi_1 f$ sends a generator $[\gamma] \in \pi_1(S^1, 1)$ to an integer multiple of $[\gamma]$. This integer, denoted $\deg f$, is called the *degree of f*.

A check is required to make sure that $\deg f$ does not depend on the choice of path, which follows from the fact that $\pi_1(S^1, 1)$ is abelian. Also, notice that $\deg f$ only depends on the homotopy class of f since $\pi_1 f$ and $\pi_1 g$ are equal as group homomorphisms whenever f and g are homotopy equivalent maps.

Example 6.3 The degree of the identity map on S^1 is 1. The degree of the map sending z to iz is also 1 since rotation by 90° is homotopic to the identity map. And for any $n \geq 1$, the degree of the map $z \mapsto z^n$ is n.

Theorem 6.7 If $f\colon S^1 \to S^1$ has degree $n \neq 1$, then f has a fixed point.

Proof. If f does not have a fixed point, define $h\colon S^1 \times I \to S^1$ by

$$h(x, t) = \frac{(1-t)fx + tx}{|(1-t)fx + tx|}$$

Then h gives a homotopy between f and id_{S^1} and so $\deg f = 1$. □

The notion of degree provides yet another application of $\pi_1 S^1 \cong \mathbb{Z}$.

The Fundamental Theorem of Algebra Every polynomial

$$pz = z^n + c_{n-1}z^{n-1} + \cdots + c_0$$

with $c_i \in \mathbb{C}$ and $n \neq 0$ has a root in $\mathbb{C}$.

Proof. Let $n \neq 0$, and suppose f does not have a root. Then

$$h(z, t) = \frac{p(tz)}{|p(tz)|}$$

defines a homotopy between $\frac{p}{|p|}$ and $\frac{c}{|c|}$, the latter being the constant map at c_0. Thus $\frac{p}{|p|}$ must have degree 0. On the other hand,

$$i(z, t) = \frac{t^n p(\frac{z}{t})}{|p(\frac{z}{t})|}$$

defines a homotopy between $\frac{p}{|p|}$ and the map $z \mapsto z^n$ which has degree n. Thus $0 = \deg \frac{p}{|p|} = n$, which is a contradiction. $\square$

With $\pi_1 S^1$ in hand, let's now turn to compute the fundamental group of other spaces. A result of Seifert and van Kampen gives us a tool for doing so.

6.7 The Seifert van Kampen Theorem

There is a common strategy employed in mathematics, which we mentioned early on in section 2.1: information about parts and how they interact is often used to obtain information about a whole. This is especially true in topology, where (as we've seen) spaces are oftentimes decomposed into open sets and information about those sets and their intersection is used to obtain information about the space. This approach is particularly valuable when computing fundamental groupoids (and fundamental groups). If a space X can be decomposed as the union of open sets U and V and if the fundamental groupoids of U, V and $U \cap V$ are known, then we can expect to understand something about the fundamental groupoid of X. As the next theorem shows, it can be understood completely—it is a colimit involving the fundamental groupoids of the spaces comprising X.

Seifert van Kampen Theorem Suppose U and V are open subsets of a topological space $X = U \cup V$. Then one has the following diagram of spaces and continuous functions:

$$\begin{array}{ccc} U \cap V & \longrightarrow & U \\ \downarrow & & \downarrow \\ V & \longrightarrow & X \end{array}$$

Applying π_1 yields a pushout diagram in the category of groupoids.

$$\begin{array}{ccc} \pi_1(U \cap V) & \longrightarrow & \pi_1 U \\ \downarrow & \ulcorner & \downarrow \\ \pi_1 V & \longrightarrow & \pi_1 X \end{array}$$

Proof. Here is the idea of the proof. If G is any groupoid fitting into a diagram as on the left,

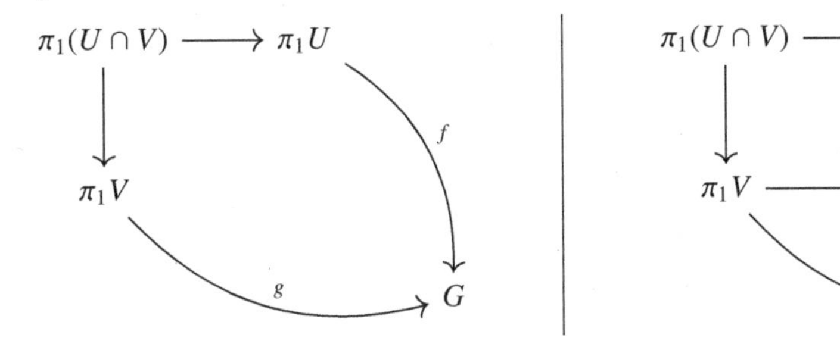

then we must construct a functor Φ completing the diagram as on the right. On objects, this is straightforward. Let $x \in X$. If $x \in U$, define $\Phi x = fx$, and if $x \in V$, define $\Phi x = gx$. If $x \in U \cap V$, these two assignments agree.

To define Φ on a homotopy class of paths from $x \to y$, choose a representative path $\gamma\colon I \to X$, and use compactness of I to subdivide the path γ into a composition of paths $\gamma_n \cdots \gamma_1$, each component of which lies in either U or V. Then, define $\Phi([\gamma])$ to be the composition of $(f \text{ or } g)[\gamma_n] \cdots (f \text{ or } g)[\gamma_2](f \text{ or } g)[\gamma_1]$ as the case may be. To see that $\Phi[\gamma]$ is well defined, let $\gamma' \simeq \gamma$, and choose a homotopy h between γ and γ'. Use compactness of $I \times I$ to subdivide the image of h into rectangles that lie entirely in U or V. The details are left as an exercise. □

The next result is an important consequence.

Proposition 6.2 Suppose U and V are open subsets of a topological space $X = U \cup V$, and suppose $x_0 \in U \cap V$. Then one has the following diagram of spaces and continuous functions:

$$\begin{array}{ccc} U \cap V & \longrightarrow & U \\ \downarrow & & \downarrow \\ V & \longrightarrow & X \end{array}$$

If $U \cap V$ is path connected, then the following diagram is a pushout in the category of groups.

$$\begin{array}{ccc} \pi_1(U \cap V, x_0) & \longrightarrow & \pi_1(U, x_0) \\ \Big\downarrow & & \Big\downarrow \\ \pi_1(V, x_0) & \longrightarrow & \pi_1(X, x_0) \end{array}$$

Proof. By the general remark following the definition of the fundamental group, the fundamental groups are equivalent as categories to the fundamental groupoids since $U \cap V$ is path connected. □

To see why $U \cap V$ must be path connected, consider the circle $X = S^1$, and let U and V be the subsets as indicated.

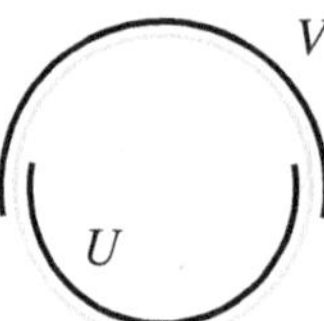

Then $U \cap V$ is homotopy equivalent to a two-point space and U and V are both contractible. Assuming Seifert van Kampen gives a supposed pushout of groups,

$$\begin{array}{ccc} 1 & \longrightarrow & 1 \\ \downarrow & \ulcorner & \downarrow \\ 1 & \longrightarrow & \pi_1(S^1, 1) \end{array}$$

But the pushout of the diagram $1 \leftarrow 1 \rightarrow 1$ is the trivial group 1, and thus $1 \cong \pi_1(S^1, 1) \cong \mathbb{Z}$, which is of course a contradiction.

Now you might be wondering, "What exactly *are* pushouts in the category Grp?" The coproduct of two groups G and H is their free product $G * H$, the group generated by the generators of both G and H with relations (equations satisfied by the generators) coming from the relations in G and H. The pushout, then, of a diagram of groups $H \leftarrow K \rightarrow G$ is a quotient of the free product such that the diagram commutes:

$$\begin{array}{ccc} K & \xrightarrow{f} & G \\ {\scriptstyle g}\downarrow & & \downarrow \\ H & \longrightarrow & (G * H)/N \end{array}$$

One concludes that N must be the (smallest) normal subgroup of $G * H$ generated by the relations $fk = gk$ for each $k \in K$. This construction is sometimes called the *amalgamated free product*.

6.7.1 Examples

Let's close by using the Seifert van Kampen theorem to compute the fundamental group of some familiar spaces.

Example 6.4 Suppose $X = S^2$ is the sphere. Let U be all of S^2 except for the point $(0, 0, 1)$, and let V be all of S^2 except for $(0, 0, -1)$. Then $U \cap V$ is homotopy equivalent to a circle, and thus its fundamental group is isomorphic to $\mathbb{Z}$. And since both U and V are contractible we obtain the diagram of groups

$$\begin{array}{ccc} \mathbb{Z} & \longrightarrow & 1 \\ \downarrow & & \downarrow \\ 1 & \longrightarrow & \pi_1(S^2, 1) \end{array}$$

The fundamental group of S^2 is then trivial.

Example 6.5 Consider the wedge product of two circles $S^1 \vee S^1$, a "figure eight." Let U and V be the indicated subspaces.

Notice that $U \cap V$ is contractible while U and V are each homotopy equivalent to a circle. Thus, the fundamental group of each is isomorphic to $\mathbb{Z}$. This is isomorphic to the free group on a single generator, so let's write $\pi_1 U \cong F\alpha$ and $\pi_1 V \cong F\beta$ where α and β are the loops generating $\pi_1 U$ and $\pi_1 V$, respectively. Then we have the diagram:

$$\begin{array}{ccc} 1 & \longrightarrow & F\alpha \\ \downarrow & & \downarrow \\ F\beta & \longrightarrow & \pi_1(S^1 \vee S^1, s_0) \end{array}$$

The fundamental group of $S^1 \vee S^1$ is therefore the free group on two generators, $F\alpha * F\beta \cong F(\alpha, \beta)$.

Example 6.6 Recall from example 1.17 that we can view the torus T as the quotient of a square with opposite sides identified:

All four corners of the square are identified to a single point, say, t_0. So let's consider the pointed torus (T, t_0). Now suppose p is any other point in T. Set $V = T \setminus \{p\}$, and let U be a "small" disc containing both p and t_0. Naming the two edges α and β, we have a situation like this:

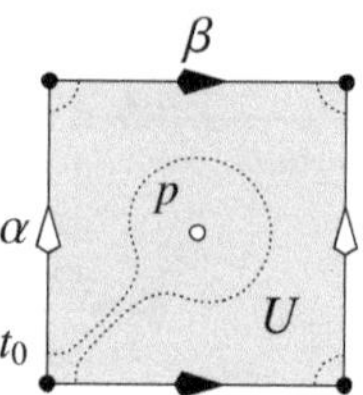

Then U is contractible. Further, V retracts onto the wedge of the loops α and β. To see this, think of removing the point p and retract the remaining gray area onto the boundary of the square. Keeping in mind that opposite sides are identified, one obtains the wedge product of the loops (α, t_0) and (β, t_0), namely, $(S_1 \vee S_1, t_0)$. Hence, by example 6.5, $\pi_1(V, t_0)$ is given by the free product on two generators which we may as well call α and β. Finally, the intersection of U and V is a punctured disc: $U \setminus \{p\}$ which retracts onto the inner "lollipop" below

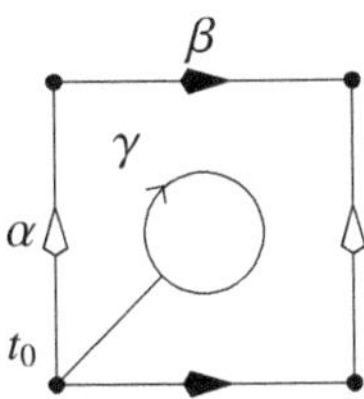

whose fundamental group is given by the free group generated by the loop γ moving from t_0 out along the diagonal, counterclockwise around the circle, and back down the diagonal to t_0. Therefore, by Seifert van Kampen,

$$\begin{array}{ccc} F\gamma & \longrightarrow & 1 \\ \downarrow & & \downarrow \\ F(\alpha,\beta) & \longrightarrow & \pi_1(T, t_0) \end{array}$$

But under the inclusion $U \cap V \to V$, the loop γ is sent to itself. And under the retraction of V onto the boundary of the square, γ maps to $\alpha\beta\alpha^{-1}\beta^{-1}$; that is, the group homomorphism $F\gamma \to F(\alpha,\beta)$ sends γ to $\alpha\beta\alpha^{-1}\beta^{-1}$. Therefore $\pi_1(T, t_0)$ is the quotient of $F(\alpha,\beta)$ generated by the relation $\alpha\beta\alpha^{-1}\beta^{-1} = 1$. This group is isomorphic to $\mathbb{Z} \times \mathbb{Z}$.

This example shows that $\pi_1(S^1 \times S^1) \cong \pi_1 S^1 \times \pi_1 S^1$, which isn't too surprising. As the categorically minded reader can check, the functor π_1 takes products to products.

Example 6.7 Referring again to example 1.17, recall that the Klein bottle K is obtained by identifying opposite sides of a square as shown.

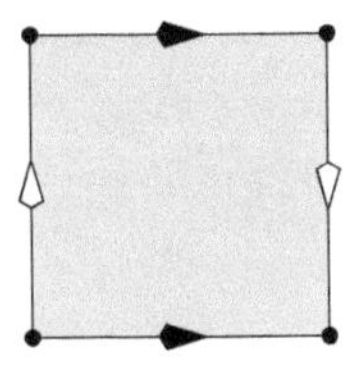

We can compute its fundamental group in the same way as for the torus. Let k_0 be the single vertex of the square, and let p be another point in K. If U is an open disc containing p and $V = K \setminus \{p\}$, then by the same arguments as in the previous example, we have the following setup

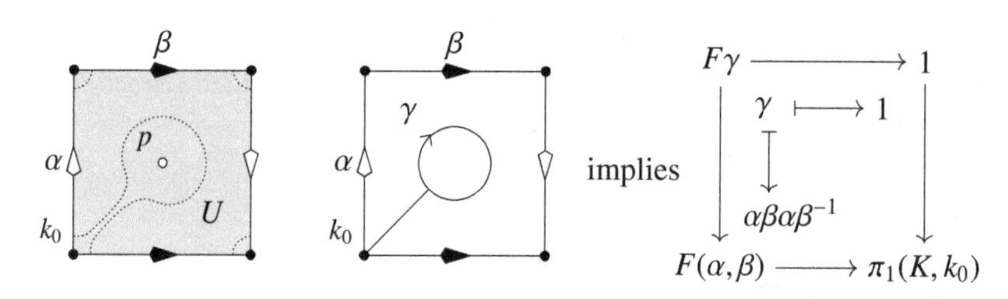

by Seifert van Kampen. The conclusion is that $\pi_1(K,k_0)$ is isomorphic to a group with a *presentation* given in two generators and one relation, namely, $\langle \alpha, \beta \mid \alpha\beta\alpha\beta^{-1} \rangle$.

But before too heartily congratulating ourselves on the calculation, recall that deciding whether a given group presentation describes the trivial group or not can't algorithmically be determined (Wikipedia, 2019). This is a variant of the word problem and is indeed equivalent to the halting problem in computing. Some wariness in working with group presentations is in order.

Exercises

1. Prove that the maps $d_0, d_1 \colon X \to X \times I$ defined by $d_0 x = (x, 0)$ and $d_1 x = (x, 1)$ and the projection $s \colon X \times I \to X$ are homotopy equivalences.

2. Prove that if $f \colon S^1 \to S^1$ satisfies $\|f(x) - x\| < 1$ for all x, then f is surjective.

3. The n-dimensional projective space is naturally pointed. Its basepoint is the class of the basepoint of S^n in the quotient $\mathbb{RP}^n \simeq S^n/\sim$, where antipodal points have been identified.

a) Prove that $\pi_1 \mathbb{RP}^2 \cong \mathbb{Z}/2\mathbb{Z}$.

b) Compute $\pi_1(\mathbb{RP}^2 \vee \mathbb{RP}^2)$.

c) Prove or disprove: $\mathbb{RP}^2 \vee \mathbb{RP}^2$ is a retract of $\mathbb{RP}^2 \times \mathbb{RP}^2$.

4. Learn what the *Eckmann-Hilton argument* is and how to use it to prove that the higher homotopy groups of a space are abelian.

5. Suppose A and Y are locally compact and Hausdorff and $f \colon A \to Y$ is a cofibration. Prove that for any space Z, the map $f^* \colon Z^Y \to Z^X$ is a fibration.

Glossary of Symbols

Symbol	Description	Page
$B(x, r)$	ball of radius $r > 0$ centered at x	2
∂	boundary	37
C	generic notation for a category	3
C^{op}	opposite category of C	6
CG	category of compactly generated spaces together with continuous maps	111
CGWH	category of compactly generated weakly Hausdorff spaces together with continuous maps; a convenient category of spaces	111
CH	category of compact Hausdorff spaces together with continuous maps	99
$\mathbb{C}$	complex numbers	22
CX	the (reduced) cone of a (pointed) space X	124
D^n	closed unit ball in $\mathbb{R}^n$	3
$\varnothing$	the empty set	1
$\twoheadrightarrow$	an epimorphism	14
$\mathbf{k}$	generic notation for a field	5
Fld	category of fields	16
Grp	category of groups	5
$\hat{f}$	shorthand for the adjunct of a map f in some adjunction	92
$\simeq$	homotopy	34
hTop	homotopy category of spaces	5
hTop_*	homotopy category of pointed spaces	121
$\mathbb{Z}$	integers: $\ldots, -2, -1, 0, 1, 2, \ldots$	22
$L \dashv R$	generic notation indicating that the functors L and R form an adjunction	92
l_p	for $1 \leq p \leq \infty$, the normed vector space of ($\mathbb{R}$-valued) sequences which converge in the p-norm	23
M_f	mapping cylinder of f	130
$\rightarrowtail$	a monomorphism	14

$\mathsf{Nat}(F, G)$	natural transformations between functors $F, G\colon \mathsf{C} \to \mathsf{D}$. *alt*: $\mathsf{D}^{\mathsf{C}}(F, G)$	12
$\mathbb{N}$	natural numbers: $0, 1, 2, \ldots$	22
P_f	mapping path space of f	130
π_n	for each $n \in \mathbb{N}$, the nth homotopy functor defined by $[S^n, -]\colon \mathsf{Top}_* \to \mathsf{Set}$	121
π_1	denotes the functor sending spaces to fundamental group(oid)s possibly relative to a subspace or a point	119, 120
$R\mathsf{Mod}$	category of modules over a ring R	5
$\mathbb{R}$	real numbers	2
$\mathbb{RP}^n$	n-dimensional real projective space	29
Set	category of sets	5
Set_*	category of pointed sets	5
ΣX	the reduced suspension of a pointed space X	124
$\operatorname{spec} R$	set of prime ideals of (a ring) R	22
S^n	n-sphere	3
SX	the suspension of a space X	125
$\mathcal{T}_x$	the open neighborhoods of x in the topology $\mathcal{T}$	2
Top	category of topological spaces	5
Top_*	category of pointed spaces	5
$\mathcal{T}$	generic notation for a topology	1
$\mathsf{Vect}_{\mathbf{k}}$	category of $\mathbf{k}$-vector spaces	5
WH	category of weakly Hausdorff spaces together with continuous maps	111
X^*	dual space of an R-module X. *alt*: $\hom(X, R)$	17

Bibliography

Arens, Richard, and James Dugundji. 1951. Topologies for function spaces. *Pacific Journal of Mathematics* 1 (1): 5–31.

Brown, Ronald. 1964. Function spaces and product topologies. *The Quarterly Journal of Mathematics* 15 (1): 238–250.

Brown, Ronald. 2006. *Topology and groupoids*, 3rd ed. BookSurge.

Cartan, Henri. 1937a. Filtres et ultrafiltres. *Comptes rendus de l'Académie des Sciences* 205: 777–779.

Cartan, Henri. 1937b. Théorie des filtres. *Comptes rendus de l'Académie des Sciences* 205: 595–598.

Chernoff, Paul R. 1992. A simple proof of Tychonoff's theorem via nets. *American Mathematical Monthly* 99 (10): 932–934.

Day, B. J., and G. M. Kelly. 1970. On topological quotient maps preserved by pullbacks or products. *Mathematical Proceedings of the Cambridge Philosophical Society* 67 (3): 553–558. doi:10.1017/S0305004100045850.

Dyson, Freeman. 2009. Birds and frogs. *Notices of the American Mathematical Society* 56: 212–223.

Eilenberg, Samuel. 1949. On the problems of topology. *Annals of Mathematics* 50 (2): 247–260.

Eilenberg, Samuel, and Saunders MacLane. 1945. Relations between homology and homotopy groups of spaces. *Annals of Mathematics* 46 (3): 480–509.

Escardó, Martín, and Reinhold Heckmann. 2002. Topologies on spaces of continuous functions. *Topology Proceedings* 26: 545–564.

Fox, Ralph H. 1945. On topologies for function spaces. *Bulletin of the American Mathematical Society* 51 (6): 429–432.

Freitas, Jorge Milhazes. 2007. An interview with F. William Lawvere - part one. *CIM Bulletin (December).* http://www.cim.pt/docs/82/pdf.

Freyd, Peter. 1969. Several new concepts: Lucid and concordant functors, pre-limits, pre-completeness, the continuous and concordant completions of categories. In *Category Theory, Homology Theory and Their Applications III*, ed. P. J. Hilton, 196–241. Springer.

Golomb, Solomon W. 1959. A connected topology for the integers. *The American Mathematical Monthly* 66 (8): 663–665.

Grothendieck, Alexander. 1997. Sketch of a programme (translation into English). In *Geometric Galois Actions, Vol. 1: Around Grothendieck's Esquisse d'un Programme*, eds. L. Schneps and P. Lochak. London Mathematical Society Lecture Notes No. 242: 243–283.

Hatcher, Allen. 2002. *Algebraic topology*. Cambridge University Press.

Hausdorff, Felix, and John R. Aumann. 1914. *Grundzüge der mengenlehre*. Veit.

Isbell, John R. 1975. Function spaces and adjoints. *Mathematica Scandinavica* 36 (2): 317–339. http://www.jstor.org/stable/24491137.

Jackson, Allyn. 1999. Interview with Henri Cartan. *Notices of the American Mathematical Society* 46 (7): 782–788.

Kadets, Mikhail Iosifovich. 1967. Proof of the topological equivalence of all separable infinite-dimensional banach spaces. *Functional Analysis and Its Applications* 1 (1): 53–62. http://dx.doi.org/10.1007/BF01075865.

Kelley, John. 1950. The Tychonoff product theorem implies the axiom of choice. *Fundamenta Mathematicae* 37 (1): 75–76.

Kelley, John. 1955. *General topology*. Van Nostrand.

Leinster, Tom. 2013. Codensity and the ultrafilter monad. *Theory and Applications of Category Theory* 28 (13): 332–370.

Lewis, Lemoine Gaunce. 1978. The stable category and generalized Thom spectra. PhD diss., University of Chicago..

Lipschutz, Seymour. 1965. *Schaum's outline of theory and problems of general topology*. McGraw-Hill.

Mac Lane, Saunders. 2013. *Categories for the working mathematician*. Vol. 5 of *Graduate Texts in Mathematics*. Springer.

Manes, E. 1969. A triple theoretic construction of compact algebras. *Seminar on Triples and Categorical Homology Theory* 80: 73–94.

Massey, William S. 1991. *A basic course in algebraic topology*. Springer.

May, J. P. 1999. *A concise course in algebraic topology*. University of Chicago Press.

May, J. P. 2000. An outline summary of basic point set topology. Miscellaneous math notes, J. P. May (website), University of Chicago. http://www.math.uchicago.edu/~may/MISC/Topology.pdf.

McCord, M. C. 1969. Classifying spaces and infinite symmetric products. *Transactions of the American Mathematical Society* 146: 273–298.

Mercer, Idris David. 2009. On Furstenberg's proof of the infinitude of primes. *The American Mathematical Monthly* 116 (4): 355–356.

Moore, Eliakim Hastings. 1915. Definition of limit in general integral analysis. *Proceedings of the National Academy of Sciences* 1 (12): 628–632.

Moore, Eliakim Hastings, and Herman Lyle Smith. 1922. A general theory of limits. *American Journal of Mathematics* 44 (2): 102–121.

Munkres, James R. 2000. *Topology*. Prentice Hall.

Nandakumar, R., and N. Ramana Rao. 2012. Fair partitions of polygons: An elementary introduction. *Proceedings—Mathematical Sciences* 122 (3): 459–467.

Render, Hermann. 1993. Nonstandard topology on function spaces with applications to hyperspaces. *Transactions of the American Mathematical Society* 336 (1): 101–119.

Riehl, E. 2014. *Categorical homotopy theory*. Cambridge University Press.

Riehl, E. 2016. *Category theory in context*, 1st ed. Dover.

Rotman, Joseph J. 1998. *An introduction to algebraic topology*. Springer.

Schechter, Eric. 1996. *Handbook of analysis and its foundations*, 1st ed. Academic Press.

Shimrat, M. 1956. Decomposition spaces and separation properties. *The Quarterly Journal of Mathematics* 7 (1): 128–129.

Spivak, David I. 2014. *Category theory for the sciences*. MIT Press.

Stacey, Andrew, David Corfield, David Roberts, Mike Shulman, Toby Bartels, Todd Trimble, and Urs Schreiber. 2019. nLab (wiki-lab). https://ncatlab.org.

Steen, Lynn Arthur, and J. Arthur Seebach. 1995. *Counterexamples in topology*. Dover.

Steenrod, Norman E. 1967. A convenient category of topological spaces. *Michigan Mathematical Journal* 14 (2): 133–152.

Strickland, Neil P. 2009. The category of CGWH spaces. Preprint, University of Sheffield. https://neil-strickland.staff.shef.ac.uk/courses/homotopy/cgwh.pdf.

Strøm, A. 1972. The homotopy category is a homotopy category. *Archiv der Mathematik* 23 (1): 435–441.

tom Dieck, Tammo. 2008. *Algebraic topology*. European Mathematical Society.

Wikipedia. 2019. Word problem for groups. Updated November 6, 2019. https://en.wikipedia.org/wiki/Word_problem_for_groups.

Wilansky, Albert. 1967. Between T_1 and T_2. *The American Mathematical Monthly* 74 (3): 261–266.

Willard, Stephen. 1970. *General topology*. Courier Corporation.

Wittgenstein, Ludwig. 1922. *Tractatus Logico-Philosophicus*. Routledge and Kegan Paul.

Ziegler, Gunter M. 2015. Cannons at sparrows. *Newsletter of the European Mathematical Society* 1 (95): 25–31.

Index